CIVIL-MILITARY RELATIONS IN

Civil-Military Relations in Perspective

Strategy, Structure and Policy

Edited by
STEPHEN J. CIMBALA
Penn State Brandywine, USA

Routledge
Taylor & Francis Group

LONDON AND NEW YORK

First published 2012 by Ashgate Publishing

Published 2016 by Routledge
2 Park Square, Milton Park, Abingdon, Oxfordshire OX14 4RN
711 Third Avenue, New York, NY 10017, USA

First issued in paperback 2016

Routledge is an imprint of the Taylor & Francis Group, an informa business

British Library Cataloguing in Publication Data
Civil-military relations in perspective : strategy,
   structure and policy.
   1. Civil-military relations--United States. 2. United
   States--Military policy. 3. National security--United
   States.
   I. Cimbala, Stephen J.
   322.5'0973-dc22

Library of Congress Cataloging-in-Publication Data
Civil-military relations in perspective : strategy, structure and policy / [edited] by Stephen J. Cimbala.
   p. cm.
   Includes bibliographical references and index.
   ISBN 978-1-4094-2978-4 (hbk)
   1. Civil-military relations. 2. Civil-military relations--United States.
   I. Cimbala, Stephen J.
   JF195.C63 2012
   322'.5--dc23

                                                                2011033437

ISBN 13: 978-1-138-26875-3 (pbk)
ISBN 13: 978-1-4094-2978-4 (hbk)

Contents

List of Figures and Tables

Figures

Tables

List of Contributors

Stephen J. Blank is Professor of Russian National Security Studies at the Strategic Studies Institute of the U.S. Army War College in Pennsylvania. Dr. Blank has been Professor of National Security Affairs at the Strategic Studies Institute since 1989. Between 1998-2001, he was Douglas MacArthur Professor of Research at the War College. He has published over 700 articles and monographs on Soviet/ Russian, U.S., Asian, and European military and foreign policies, testified frequently before Congress on Russia, China, and Central Asia, consulted for the CIA, major think tanks and foundations, chaired major international conferences in the U.S.A. and abroad in Florence, Prague, and London, and has been a commentator on foreign affairs in the media in the United States and abroad. He has also advised major corporations on investing in Russia and is a consultant for the Gerson Lehrmann Group. Dr. Blank has published or edited 15 books focusing on Russian foreign, energy, and military policies and on international security in Eurasia. His most recent book is *Russo-Chinese Energy Relations: Politics in Command* (2006). He has also published *Natural Allies?: Regional Security in Asia and Prospects for Indo-American Strategic Cooperation* (2005), in addition to other books and monographs.

Stephen J. Cimbala is Distinguished Professor of Political Science at Penn State Brandywine and has contributed to the literature of national security studies for many years. Dr. Cimbala is an award-winning Penn State teacher, has consulted for a number of U.S. government agencies and contractors, and served as a university administrator for almost a decade. Cimbala's recent books include *Multinational Military Intervention* (2010) with Peter K. Forster and *The George W. Bush Defense Program* (2010), for which he served as contributing editor.

Damon Coletta is Professor of Political Science at the U.S. Air Force Academy. He was first trained as an electrical engineer, working on electromagnetic compatibility issues for avionics on the C-17 aircraft in the early 1990s. Dr. Coletta earned a master's degree in Public Policy from the Kennedy School of Government, Harvard University, specializing in Science and Technology Policy. He has performed research for the American Association for the Advancement of Science, the American Enterprise Institute, the National Aeronautics and Space Administration, the Institute for Defense Analyses, and RAND Corporation. His work has appeared in *International Organization, Contemporary Security Policy, Armed Forces & Society, International Studies Review,* and *Foreign Policy Analysis.* He co-edited the eighth edition of *American Defense Policy* (Johns Hopkins University Press, 2005) and *Space and Defense Policy* (Routledge, 2009). Damon also authored *Trusted Guardian: Information Sharing and the Future of the Atlantic Alliance* (Ashgate, 2008). Dr. Coletta received his Ph.D. in political science from Duke University.

Edward Cox is Assistant Professor of American Politics at the U.S. Military Academy, West Point. He has served in various command and staff positions in combat units for 12 years, including two years conducting combat operations in Iraq. He holds a bachelor's degree in American politics from the U.S. Military Academy and master's degrees in public administration and international relations from Syracuse University. He has been published in *Military Review* and the Association of the U.S. Army's Land Warfare Series, and his first book, a biography of Major General Fox Conner, was published by New Forums Press, 2011.

Lester W. Grau is a senior analyst for the Foreign Military Studies Office at Fort Leavenworth, Kansas. He retired from the Army in 1992 after having served in Vietnam, Korea, and Europe, including a posting in Moscow. Dr. Grau has published over 100 articles and several books, including *The Bear Went Over the Mountain* and other definitive studies on the Soviet war in Afghanistan. He holds a B.A. and an M.A. in international relations and a Ph.D. in military history.

Dale R. Herspring, a University Distinguished Professor of Political Science at Kansas State University and a member of the Council on Foreign Relations, is a retired U.S. diplomat and Navy captain. Dr. Herspring is the author/editor of more than 90 articles and 14 books. His most current work is to be entitled *Military Culture and Civil-Military Relations: The American, Russian, German and Canadian Experiences*. His recently published scholarly books include: *After Putin's Russia: Past Imperfect, Future Uncertain* (Fourth Edition, 2010), with Stephen K. Wegren; *Rumsfeld's Wars: The Arrogance of Power* (2008); and *The Kremlin and the High Command: Presidential Impact on the Russian Military from Gorbachev to Putin* (2006).

Jacob W. Kipp retired from federal service in September 2009 and is currently an Adjunct Professor at the University of Kansas and a weekly columnist on Eurasian security for the Jamestown Foundation. Dr. Kipp received his Ph.D. in Russian history from Pennsylvania State University and has published extensively on Russian and Soviet naval and military history.

Adam Lowther (B.A., Arizona State University; M.A., Arizona State University; Ph.D., University of Alabama) is a Defense Analyst at the Air Force Research Institute (AFRI), Maxwell AFB, AL. His principal research interests include deterrence, nuclear weapons policy, and terrorism. Dr. Lowther is the editor of *Terrorism's Unanswered Questions* and the author of *Americans and Asymmetric Conflict: Lebanon, Somalia, and Afghanistan*. He has published in the *New*

York Times, Boston Globe, Joint Forces Quarterly, Strategic Studies Quarterly and in other publications. Prior to joining AFRI, Dr. Lowther was an Assistant Professor of Political Science at Arkansas Tech University and Columbus State University, where he taught courses in international relations, political economy, security studies, and comparative politics. Early in his career, Dr. Lowther served in the U.S. Navy aboard the U.S.S. *Ramage* (DDG-61). He also spent time at CINCUSNAVEUR London, and with NMCB-17.

Kent W. Park is Instructor of American Politics at the United States Military Academy, West Point, New York. Most recently, he served as the commander for a Stryker Infantry Company from Fort Wainwright, Arkansas. Major Park was commissioned as a U.S. Infantry officer from West Point with a bachelor's degree in International Relations. He also holds a master's degree in public policy from the John F. Kennedy School of Government, Harvard University. He has published and presented his research in various forums, including the Kennedy School Review, the Center for Public Leadership, the Association of the U.S. Army's Land Warfare Series, and in the Boston Globe.

Gary Schaub, Jr. (B.S., Carnegie Mellon University; M.A., University of Illinois at Urbana; Ph.D., University of Pittsburgh) is an Assistant Professor of Strategy and Leadership at the Air War College, Maxwell AFB, AL. His principal research interests include coercion, civil-military relations, and political psychology. Dr. Schaub has published in *The New York Times*, *The Washington Times*, *The Pittsburgh Post-Gazette*, *The Air Force Times*, *Parameters*, *Strategic Studies Quarterly*, *Defence Studies*, *Political Psychology*, and elsewhere. He is a contributing editor to *Strategic Studies Quarterly* and has consulted Joint Forces Command, Strategic Command, and European Command. Prior to joining the Air War College faculty, Dr. Schaub was a Visiting Assistant Professor at the School of Advanced Air and Space Studies.

Rachel M. Sondheimer is Assistant Professor in the American Politics, Policy and Strategy stem of the Department of Social Sciences at the United States Military Academy, West Point. Dr. Sondheimer's work has been published in the *American Journal of Political Science* and numerous edited volumes. Her recent research includes analysis of political behavior and beliefs of active members of the military and military communities with a focus on the impact of this behavior on civil-military relations. Dr. Sondheimer received her Ph.D. in political science at Yale University in 2006 and her undergraduate degree in government from Dartmouth College in 2001.

Milan Vego was born in Capljina, Bosnia and Herzegovina, leaving former Yugoslavia in July 1973, he received political asylum in February, 1976 in the United States, and has been a U.S. citizen since April 1984. Dr. Vego is a graduate of the former Yugoslav Naval Academy (Class of 1961); B.A. in Modern History,

University of Belgrade (1970); M.A. in Modern History, University of Belgrade (1973) (U.S. History/Latin American History); Master Mariner's License (1973; Ph.D. in European History, the George Washington University (1981). He served as a Line Officer in the former Yugoslav Navy (1961-1973); Third and Second Mate (Deck) on board West German ships (1973-1976); freelance writer (1979-1991); Adjunct Professor, the George Washington University (1984); Visiting Fellow, Center for Naval Analyses (CNA), Alexandria, Virginia (1985-1987); Research Fellow, Foreign Military Studies Office (formerly Soviet Army Studies Office, U.S. Combined Arms Center, Fort Leavenworth, Kansas (1987-1989); Adjunct Professor at the Defense Intelligence College, DIA (1985-1991) and at the War Gaming and Simulation Center, National Defense University (1988-1991). Since 1991, he has been Professor of Operations, JMO Department, U.S. Naval War College, Newport, Rhode Island, and a tenured professor since 2001. Dr. Vego has published eight books and numerous articles in professional journals.

C. Dale Walton is a lecturer in International Relations and Strategic Studies at the University of Reading, UK, and he received his Ph.D. from the University of Hull. Prior to coming to the University of Reading, Dr. Walton taught in the Defense and Strategic Studies Department at Missouri State University. His research interests include geopolitics and the changing geostrategic environment, terrorism, and U.S. military-strategic history. He is the author of *Geopolitics and the Great Powers in the Twenty-first Century* (2007) and *The Myth of Inevitable U.S. Defeat in Vietnam* (2002).

John Allen ("Jay") Williams is Professor of Political Science at Loyola University Chicago. He is Chair and President of the Inter-University Seminar on Armed Forces and Society (www.iusafs.org) and editor of the *National Strategy Forum Review* (www.nationalstrategy.com). He is on the Board of Directors of the Pritzker Military Library (www.pritzkermilitarylibrary.org) and is co-organizer and former President of the International Security and Arms Control Section of the American Political Science Association. His academic degrees are from Grinnell College (B.A.) and the University of Pennsylvania (M.A. and Ph.D.). His writings include works on professional military education, civil-military relations, military culture, military professionalism and leadership, personnel issues, military strategy, military forces and missions, catastrophic terrorism, defense organization, and strategic policy. Dr. Williams' latest book is *U.S. National Security: Policymakers, Processes, and Politics* (with Sam C. Sarkesian and Stephen J. Cimbala). Previous books include *The Postmodern Military: Armed Forces After the Cold War* (with Charles C. Moskos and David R. Segal), *Soldiers, Society, and National Security* (with Sam C. Sarkesian and Fred B. Bryant), *The U.S. Army in a New Security Era* (with Sam C. Sarkesian), and *U.S. National Security Policy and Strategy, 1987-1995: Documents and Policy Proposals* (with Robert A. Vitas). Dr. Williams retired as a Captain in the U.S. Naval Reserve with 30 years of commissioned service. His personal awards include the Legion of Merit, the Meritorious Service Medal (two awards) and the Navy and Marine Corps Achievement Medal.

Isaiah Wilson III is an Associate Professor and Director of American Politics, Policy and Strategy with the Department of Social Sciences, U.S. Military Academy at West Point. He holds a B.S. in International Relations (USMA), and an M.P.A., M.A. and Ph.D. from Cornell University, and two M.M.A.S degrees from the U.S. Army Command and General Staff College and the School of Advanced Military Studies. Dr. Wilson is a recognized expert on theater operations planning and a member of the Council on Foreign Relations. He is the author of the book, *Thinking Beyond War: Civil-Military Relations and Why America Fails to Win the Peace* (Palgrave Macmillan, 2008).

Introduction

Stephen J. Cimbala

Civil-military relations in the United States and in other countries reflect the professional orientations and thinking of military officers, the decision-making process, as among defense bureaucracies and other institutional players in government, and the political outcomes that result from international relations and their feedback into foreign policy decisions. In short: civil-military relations constitute an expansive subject matter, with indistinct boundaries between it and national security studies, defense and foreign policy making, military history or other sub-disciplines of interest to persons in the armed forces, government and academia. In times past, civil-military relations were thought to have a center of gravity in the legal or institutional relations between the strictly "civilian" versus "military" spheres, but modern and postmodern studies have regarded this bimodal perspective as too simplistic. It is now the apparent consensus that civil and military responsibilities at least in developed countries have considerable overlap, and, therefore, shared responsibility for success or failure in war or peace.

The study of civil-military relations co-evolves with developments in the art of war in addition to social and cultural changes. As Keith F. Otterbein has noted, war and military organizations "are as important as kinship and the family, religious practices and practitioners, and the economy and modes of exchange to understanding a particular society."[1] Of the art of war, it may be said that the essential nature of war is unchanging, although the character of war changes with developments in economics, technology, culture, society, and especially, politics.[2] It may also be said that since war is socially, culturally, economically, technologically and politically determined, so, too, are civil-military relations. With respect to their lasting impacts on civil-military relations, however, not all variables are equal. Defeat and victory in past wars hold "lessons learned" for the organization, training, and leadership of armed forces in future war—and for the political leadership of defeated or victorious states and non-state actors as well. If democracies are less warlike than autocratic states, as argued by some theorists, this may have to do with the greater willingness of the former to subject prior

---

1   Keith F. Otterbein, *The Anthropology of War* (Long Grove, IL: Waveland Press, 2009), p. 3.

2   Relative to the study of civil-military relations, this point is emphasized in Mackubin Thomas Owens, *U.S. Civil-Military Relations after 9/11: Renegotiating the Civil-Military Bargain* (New York: Continuum Publishing Group, 2011), especially p. 140.

wartime experiences to critical audit and examination and to hold public officials accountable to the electorate for apparent failure.

The co-evolution of U.S. civil-military relations with the growing responsibilities assumed by American foreign policy since World War II has not been coincidental. The Cold War American Presidents were accorded unprecedented power in peacetime to manage the armed forces and, as well, a burgeoning national security establishment. President Eisenhower's farewell warning about the dangers of an emerging "military industrial complex" seems almost quaint by later Cold War and post-Cold War standards. With the end of the Cold War and the demise of the Soviet Union, the United States found itself described in the 1990s and since as a singular military superpower. The United States accepted greater responsibility for world order and commitments to multinational peace and stability operations in Bosnia, Kosovo and elsewhere during the final decade of the twentieth century. In the first decade of the twenty-first century, the attacks of 9-11 drew American military forces into Afghanistan and, indirectly, into Iraq, whose democratization was described by the George W. Bush administration as a necessary step in the war on terror. According to one noted military historian, post-Cold War success in U.S. military operations may have unbalanced the relationship between civilian and military participants in the decision-making process:

> The end of the Cold War and an operational tour de force in the first Persian Gulf War cemented the military's position as the public's most trusted and esteemed institution. During the Clinton administration, the military leadership had a virtual veto over military policy, particularly the terms and conditions of interventions overseas. The power of the military has waxed and waned since the 1940s, but not a single secretary of defense has entered office trusting the armed forces to comply faithfully with his priorities rather than their own.[3]

The Persian Gulf War of 1991 (coalition operation Desert Storm) showed that the United States had established primacy in the conduct of information-based warfare, especially in the application of long-range precision strike, C4ISR (command, control, communications, computers, intelligence, surveillance and reconnaissance), stealth, and other advanced technologies to conventional war fighting. Therefore it followed, according to the paradoxical logic of strategy, that future enemies of the United States would emphasize asymmetrical approaches, including unconventional or irregular warfare of various kinds. Insurgency and terrorism soon moved to the top of U.S. threat assessments, in addition to the various by-products of failed states that included sectarian violence, human and drug trafficking, and spreading social, economic and environmental chaos. As a result, the U.S. found itself for two decades using military force not only for the

---

3   Richard H. Kohn, "Coming Soon: A Crisis in Civil-Military Relations," *World Affairs*, Winter 2008, http://www.worldaffairsjournal.org/articles/2008-Winter/full-civil-military.html.

obvious purpose of battle, but also to support the reconstruction of societies and the reestablishment of governments. These post-conflict security and stability operations created challenges to sort out various aspects of U.S. and allied civil-military relations "on the ground": and, as well, among those security forces both military and civil that were being rebuilt in states with deposed regimes, as in Iraq and Afghanistan.[4]

In the post-Cold War world, U.S. and other civil-military relations would also involve issues of domestic as well as international security. Among these domestic security issues would be the balance of power among competing military and intelligence bureaucracies. Russia's security establishment in the 1990s was in a declared state of democratization and transparency, but in actual fact, the securitization of financial power and the commingling of private and public sector oligarchies led post-Soviet Russia into an unstable relationship between military and political power. Military reform was resisted by the Ministry of Defense and the General Staff throughout the years of Yeltsin's presidency and even during Putin's more defense-minded years in the same position. The Medvedev-Putin "tandem" committed itself to serious reform in the wake of military operational embarrassments during Russia's August, 2008 war against Georgia. Questions remained whether Russia could afford to replace as many conscripts with voluntary contract soldiers as planned, whether Russia could modernize both nuclear and conventional forces, and whether the Russian military would have first or later call upon scarce resources compared to internal security troops and intelligence organs.[5]

In turn, Russia's relations with the U.S. and NATO Europe would influence its threat perceptions and, derivatively, the balance of power among its military and other security bureaucracies. The Obama administration argued for a "reset" of relations with Russia, of which the centerpiece and starting point was the agreed New START treaty of 2010. New START was seen by its proponents as the beginning of a process of gradually improving U.S.-Russian political relations, but this favorable outcome was neither determined nor obvious. Russia, relative to its international

---

4 According to Andrew J. Bacevich, so-called long wars or protracted conflicts are antithetical to the values of democratic government, including "a code of military conduct that honors the principle of civilian control while keeping the officer corps free from the taint of politics." See Bacevich, "Endless war, a recipe for four-star arrogance," *Washington Post*, June 27, 2010, p. B01, http://www.washingtonpost.com/wp-dyn/content/article/2010/06/25/AR2010062502160_pf/

5 For pertinent analysis and commentary, see: Dale R. Herspring, "Is Military Reform in Russia for 'Real'? Yes, But," ch. 2 in *The Russian Military Today and Tomorrow: Essays in Memory of Mary Fitzgerald*, edited by Stephen J. Blank and Richard Weitz (Carlisle, PA: Strategic Studies Institute, U.S. Army War College, July 2010), pp. 151-91; Herspring, "Putin, Medvedev and the Russian Military," ch. 12 in *After Putin's Russia: Past Imperfect, Future Uncertain*, Fourth Edition, edited by Stephen K. Wegren and Dale R. Herspring (Lanham, MD: Rowman and Littlefield, 2010), pp. 265-89; and Olga Oliker, Keith Crane, Lowell H. Schwartz and Catherine Yusupov, *Russian Foreign Policy: Sources and Implications* (Santa Monica, CA: RAND Corporation, 2009), esp. pp. 139-74.

threat assessments, compartmentalized its issue areas as among economics, energy security, nuclear arms control and internal security, among others. The U.S. and Russia had overlapping and potentially favorable interests in pacifying Afghanistan and diminishing the significance of insurgency and terrorism in Central and South Asia and in the Caucasus. However, Russia has its own interest in Central Asia and its concerns about growing Chinese influence there, in addition to related "known unknowns" about China's Far Eastern objectives and military preparedness. Russia's policy-making process under Putin and Medvedev struggled to resolve upon a more rational geostrategic map underlying its threat assessments. Instead, in Russian policy statements and in official military doctrine, the U.S. and NATO have remained the principal threats of choice—although whether "threats" or "dangers" according to official rhetoric was an obviously debated point. As Stephen J. Blank has noted, with respect to Russia's civil-military relations and decision-making process with respect to national security policy and defense doctrine:

> Given the absence of the rule of law in the government and state, it is hardly surprising that policymaking remains personalized, haphazard, fragmented, and subject to endless and often inconclusive struggles. Neither should we be surprised that the Russian state is deficient in the means of conducting a true national strategy.[6]

If Russia's reaction to 9-11 was ambivalent in its embrace of the U.S. and NATO as security partners, one reason was the George W. Bush administration's assertion of U.S. options for preemptive attack against terrorists or states that sponsored terrorism and the related designation of Iraq, Iran and North Korea as an "axis of evil." After the U.S. and allied invasion of Iraq to depose the Saddam Hussein regime in 2003, over Russia's opposition and without the support of the UN Security Council, U.S.-Russian relations cooled considerably for the remainder of Putin and Bush's terms in office. The George W. Bush national security strategy not only vexed Russians, but it also created new issues for civil-military relations. Presidential candidate George W. Bush in 2000 declared an aversion to the use of U.S. armed forces in nation building, as in the 1990s. But Bush soon found himself required to sustain simultaneous post-conflict stability and reconstruction operations in Iraq and Afghanistan. Bush's ebullient Defense Secretary Donald Rumsfeld first denied the existence of an insurgency in Iraq, and things drifted from bad to worse until his replacement in the fall of 2006 and the arrival in January, 2007 of General David Petraeus in command of U.S. forces in Iraq. Prior to assuming command in Iraq, Petraeus had led army and inter-service reviews of U.S. counterinsurgency doctrine, and he applied the lessons learned from these studies in Iraq, apparently to considerable success in 2007-2008.

---

6   Stephen J. Blank, "No Need to Threaten Us, We Are Frightened of Ourselves," Russia's Blueprint for a Police State, The New Security Strategy," ch. 1 in *The Russian Military Today and Tomorrow*, edited by Blank and Weitz, pp. 19-149, citation pp. 23-4.

However, the debate over U.S. counterinsurgency doctrine was not only a matter of military strategy but also one of high politics. When the Obama policy review of 2009 sought to define the administration's future posture on Afghanistan, it became clear that the U.S. military leadership was of more than one mind about the priority of counterinsurgency and other irregular wars in Pentagon threat assessments.[7] Critics of counterinsurgency doctrine labeled the aficionados of counterinsurgency as "COINdinistas" and argued that they constituted a cult mentality that could lead astray force planning and doctrine. According to critics, U.S. armed forces trained and tasked primarily for COIN might lose their fighting edge and doctrinal clarity with respect to performing in more conventional conflicts. Critics of the COIN critics responded that it was not an either-or but a both-and: the U.S. must be prepared for conventional and unconventional conflicts.[8] However, the prospect of additional Iraqs or Afghanistans did not enamor itself to the American public nor to the U.S. Congress as the mid-term elections of 2010 approached. In addition, the Pentagon projected post-2010 downsized budgets that might preclude "having it all" across the entire spectrum of military conflict.

The implications for U.S. civil-military relations, with respect to preference for COIN-centric doctrine and training, as opposed to more conventional emphases, lay partly in the making of military careers and, especially, in the definition of unified or specified commands responsible for peacetime and wartime military performance. Even in fat budgetary years there are only so many billets for promotion to general officer or flag rank, and these promotions must be distributed across services and within competing service bureaucracies. As well, informal networks, among officers who have served together in key assignments or groomed one another's protégé's for eventual higher rank, exert considerable force within and outside the Pentagon. These professional and interpersonal networks are sometimes characterized by strongly felt and communally shared political views (the "fighter mafia" or the "bomber barons"). In addition, these intra-service or multi-service professional networks have allies and supporters in the broader policy community, among members of Congress and elsewhere in the Executive branch, and not infrequently simpatico and support from influential think tanks and media. Examples in the case of the U.S. would be the highly influential partisans of "fourth generation warfare" and "network-centric warfare" respectively, with different Pentagon and exterior power bases.

---

7 For an analysis, with respect to the Pentagon's Congressionally-mandated Quadrennial Defense Review (QDR), see Roy Godson and Richard H. Shultz, Jr., "A QDR for All Seasons? The Pentagon Is Not Preparing for the Most Likely Conflicts," *Joint Force Quarterly*, 59 (4th Quarter 2010), pp. 52-6.

8 Expert assessment of this debate and prominent schools of thought is provided in William Flavin, *Finding the Balance: U.S. Military and Future Operations* (Carlisle, PA: Peacekeeping and Stability Operations Institute [PKSOI], U.S. Army War College, March 2011).

It would be remiss to give the impression that civil-military relations are a subject matter only of concern to policy-making elites or others with exceptional political influence. In democratic societies, public opinion on military and other matters provides a larger context within which elites must decide and act. In this regard, the U.S. military finds itself held in higher regard than almost any other American profession. On the other hand, the U.S. armed forces are based entirely on voluntary recruitment, obviously efficient in peacetime but controversial in time of war. In wartime the question of shared sacrifice is raised in the body politic and that issue becomes more salient as the war becomes more protracted and the casualties increase. The American public has shown little aversion to casualties in war provided the cause is just and popular and the administration has an apparent strategy for prevailing in good time. On the other hand, the U.S public turns quickly against policy makers who wander into quagmires or launch military operations for ambiguous causes.

The immediately preceding discussion is a reminder that civil-military relations are an important aspect of strategy, that is, of strategy making. Colin Gray's theoretical model of strategy as a metaphorical "bridge" between the aims or ends of policy and the ways and means of military action invites civil-military relations into the discussion of strategy. Civil-military relations are important at each of the three crossroads of aims and arms: policy, strategy, and the threat or use of force in action. In the U.S., post-World War II studies of civil-military relations have included landmark contributions by Samuel Huntington, Morris Janowitz, Peter Feaver, Eliot Cohen, Sam Sarkesian, Mackubin Thomas Owens and others. Huntington's highly influential model of "subjective" versus "objective" civilian control over the armed forces has served as a focal point of departure, even for those in disagreement with his approach.[9] Janowitz's "constabulary" concept, Feaver's analyses of the civil-military "problematique" and civilian control of nuclear weapons, Cohen's "unequal dialogue" between civil and military power, Sarkesian's "constructive military engagement" between politicians and soldiers, and Owens' perspective of "renegotiating the civil-military bargain" have all added important insights to Huntington's approach but not entirely superseded it.[10]

---

9   Samuel P. Huntington, *The Soldier and the State: The Theory and Politics of Civil-Military Relations* (Cambridge, MA: Harvard University Press, 1957) and (New York: Vintage Books, 1964).

10   Morris Janowitz, *The Professional Soldier: A Social and Political Portrait* (New York: The Free Press, 1960, 1964); Peter Douglas Feaver, *Guarding the Guardians: Civilian Control of Nuclear Weapons in the United States* (Ithaca, NY: Cornell University Press, 1992); Eliot A. Cohen, *Supreme Command: Soldiers, Statesmen, and Leadership in Wartime* (New York: Free Press, 2002); Sam C. Sarkesian and Robert E. Connor, Jr., *The U.S. Military Profession into the Twenty-First Century: War, Peace and Politics* (New York: Routledge, 2006), Second Edition; and Owens, *U.S. Civil-Military Relations after 9/11*, previously cited. These examples do not exhaust the contributions of these authors, nor enumerate those of others equally distinguished (for example, Michael Desch, Charles Moskos, and John Allen Williams).

The study of strategy should help to educate policy makers about the uses and abuses of military art, but the practice of strategy to good effect cannot be guaranteed by improving the military education of politicians (or the political savvy of military officers). Doing strategy effectively requires a process that is both disciplined and flexible with respect to the interactions between warlord and politician.[11] Examples of dysfunctional relationships fill the pages of military histories. In democratic societies, politicians from both executive and legislative branches necessarily have the last word over policy options and the resources to support them. Thus, in the United States, the armed forces answer to a civilian Secretary of Defense and President as Commander in Chief. In addition, the U.S. Congress holds the power of the purse over the military and other bureaucracies. Getting coherent strategy out of this and other democratic systems is not impossible, but the appearances cast by the work in progress are not for the squeamish. Democracies have a talent for ending up with the "right" decision in the "wrong" manner, at least from the standpoint of theory as opposed to practice. One reason that democracies appear to get better results in strategy and in civil-military relations than they deserve is that they are more explicitly self-correcting of recognized bad habits.

The period since the end of the Cold War has been marked by increasing numbers of so-called irregular or unconventional conflicts. Many of these are based in failing or failed states and are the cross products of disintegrating political legitimacy combined with economic distress, ethno-nationalist and tribal strife, and ecological catastrophe. For outside powers attempting to mediate or otherwise conduct peacekeeping or post-conflict operations in these circumstances, civil-military relations take on new challenges. Armed forces must work with NGOs, local governments, tribes, militias and religious factions, among other players in the field of broken dreams. The reassembly of disaggregated states involves a patient reconstruction of their societies and a tamping down of the hostility between or among disparate cultures; that is, rethinking our approach to "war amongst the people" in Rupert Smith's characterization. More and more, we are learning that we cannot kill our way out of these complex contingency operations or amorphous wars. Civilian agencies are as important as military operations in these milieus for the effective application of strategy. But, at least in the American case, civil agencies have less of a deployable "field capability" than do military departments. Soldiers even in large number are inadequate substitutes for courts of law, entrepreneurs, engineers, local police, and "ground truth" intelligence that must come from locals.

David Kilcullen's excellent book on counterinsurgency demonstrates that the U.S. and allied NATO countries are smarter about effective strategy making in complex

---

11   My discussion of the relationship between civil-military relations and strategy benefits considerably from Colin S. Gray, *The Strategy Bridge: Theory for Practice* (Oxford: Oxford University Press, 2010), especially pp. 149-57. Gray, however, is not responsible for arguments here.

contingency operations than they used to be.[12] The recognition that the "center of gravity" in counterinsurgency is the protection of the general population, together with the separation of noncombatant civilians from support for the insurgents, is now almost canonical in American and other military circles. Nevertheless, U.S. and allied involvement in these situations is apt to require protracted deployment and continuing frustration over the course of a decade or longer. Counterinsurgency is a competition in civil and military patience as much as anything else. Over-resourcing the problem or turning up the rhetoric cannot compensate for lack of political perseverance or military steadfastness. It is all very well to observe that the U.S. might have done better in Iraq after "Mission Accomplished" had it deployed larger numbers of troops from the outset. And it may be equally valid to suggest that U.S. "influence operations" in Iraq improved gradually only after painful learning experience. Nevertheless, mass and Madison Avenue serve as icings on the proverbial cake of counterinsurgency strategy. Leaders and publics alike must be educated for the long war and the ways and means that follow from that recognition.

Since all politics is about power and influence, variously defined, the study of civil-military relations partakes of politics in obvious and less apparent ways. Politics establishes the context within which geostrategic, including military, and foreign policy decisions can be made. A state can attain temporary success in the performance of operational art or tactics without having demonstrated competence in grand strategy and high policy. As examples, the Japanese empire and the Third Reich were able to run the table for a time against their World War II enemies, based on superior tactical performance and operational art. However, neither Japan nor Nazi Germany connected its early tactical success to a convincing theory of victory embodied in coherent policy and strategy for a protracted major coalition war. The lacunae in German and Japanese planning for and during World War II had something to do with their civil-military relations, contributing to a noticeable deficit in policy and strategy. In the case of Japan, Admiral Yamomoto Isoroku was virtually alone in warning against an extended war with America. In Germany, Hitler's answer to the unexpected resistance of Britain to defeat in a short war was to invade Russia (!)—double-crossing his former comrades in arms in Moscow, and ensuring that a winning coalition would be in place against Germany once the U.S. became a belligerent.

An interesting question with respect to civil-military relations is whether modernizing autocracies with an Eastern or Middle Eastern way of war can benefit from the experience of Western militaries and civil-military relations. Can China's rising star be propelled by a civil-military relationship that avoids the worst of the Soviet system, enabling professional military competence within the larger communist party power structure and rule? Can India's emergence as a military great power, at least regionally, benefit from the study of past successes and failures by Western democracies in controlling their armed forces and in

---

12 David Kilcullen, *Counterinsurgency* (Oxford: Oxford University Press, 2010), especially chs 1-2 and 6.

the formulation of policy and strategy? Will the regime in post-Soviet Russia work out a relationship with its reforming military that allows a transition to information age competence, or will Russia remain in retro with reliance on a deficient pool of conscripts, a Soviet-style view of military art, and a propensity for military threat assessments that remains mired in the Cold War (or earlier) past? And what futures portend for civil-military relations in Iran, Iraq, Saudi Arabia and other influential regional actors in the Arab and Islamic worlds?

Modern Western militaries had more or less resolved the relationship between church and state, between scepter and miter, before embarking on the industrial and later revolutions in military affairs. But some Middle Eastern and South Asian armed forces will be tasked to formulate military strategy and doctrine within a political context highly embedded in religious symbolism and, in some cases, involving the clergy in control of organs of state. A return of the Ottoman Empire is improbable, but an arc of uncertainty about civil-military relations extends across North Africa through the Eastern Mediterranean—Levant, Turkey, the Arabian Peninsula, Persia and Mesopotamia, former Soviet Central Asia, Afghanistan and Pakistan. Secular governments in some of these regions are under pressure for Islamicization of their politics, including the politicization of their security organs and militaries. Pakistan already finds itself a divided house marked by political conflict between secular pluralists and dissident Islamicists of various types, and these conflicting tendencies play out within that country's armed forces and intelligence bureaucracies. Pakistan also possesses nuclear weapons and, however ambivalent about the Taliban in Afghanistan from the U.S. and NATO perspective, cannot be avoided as the proving ground for success or failure in stabilizing a Karzai regime in Afghanistan.

And speaking of nuclear weapons, the possible spread of nuclear weapons among more states in Asia and-or the Middle East raises a number of issues for civil-military relations. Space does not permit an extended discussion, but the short form of nuclear history is as follows. Through protracted trial and error, the U.S., Soviet and post-Soviet Russia, and other twentieth century nuclear powers learned important lessons about the operation, management and control of nuclear forces. Speaking broadly, nuclear weapons, launchers, and infrastructure required specialized chains of command and hierarchies of control, with "fail safe" protocols both technical and procedural to ensure against (1) the possibility of an accidental or mistaken launch of a nuclear first strike or first use; and (2) the failure of nuclear forces to carry out a successful retaliation against an enemy first strike, due to technical malfunction or flawed decision procedures. The ingredients of failure for possibility number (1), as above, included military usurpation of civilian command over the nuclear launch decision during a crisis or coup attempt. The constituent elements of failure for possibility number (2), as above, included decapitation of the political or military chains of command and disruption of procedures for delegation of authority to surviving commanders.

Given the consequences of a U.S.-Soviet nuclear blowout on account of a failure of deterrence during the High Cold War, John Keegan is probably correct to refer to the tasks of nuclear-age heads of state and government and force commanders as "post-heroic" in their mission and professional orientation. They and their states are denied an honorable endgame of prevailing in battle at an acceptable cost, relative to the possible outcomes of conventional wars. The realization that nuclear strategy is therefore primarily or exclusively about the avoidance of war, instead of being about the combative use of nuclear weapons to strategic effect, may make for a controlled nuclear proliferation in which deterrence remains uncertain, but also untested in practice. However, given history's propensity for wars driven by "fear, honor and interest" as Thucydides noted, reliance on deterrence in the face of extensive nuclear weapons spread could be the equivalent of wishful thinking or gallows humor.

Tutorials in civil-military relations for emerging nuclear weapons states, offered by those already members of the nuclear club, may be a "necessary evil" in order to avoid technical or political failure of nuclear command and control. Some evidence of success in this regard is apparent in Pakistan's recent reorganization of its nuclear security arrangements, doubtless with the blessing of U.S. political and military leaders and the backing of U.S. nuclear expertise. Improving civil-military relations within emerging or nascent nuclear powers implies greater clarity about "who" can enable a nuclear launch, under "what" circumstances and with "which" checks and balances, and "how" the various nuclear weapons and launchers are stored in peacetime and made ready during crises.

Nuclear weapons are another reminder of the significance of the ethical dimension of civil-military relations. It is not accidental that the motto of West Point includes duty, honor and country. A sense of honor is more important to the military than it is to almost any other profession, save, perhaps, the clergy. The officer is charged with being a gentleman (or gentlewoman), but no longer on account of the correlation between military leadership and aristocracy. Instead, the officer-as-gentleman points to the nobility of purpose and the strength of character, including performance in combat, that separate the military professional from most of his or her civilian counterparts. Although, for example, academics have their honor codes of sorts (against plagiarism, for example), professors are infrequently required to be shot at as a condition for promotion and tenure. The Duke of Wellington captures this nobility of purpose and strength of character in his official Waterloo dispatch of June 19, 1815 in which he commends various officers (including wounded and killed in action) and notes:

> I propose to move, this morning, upon Nivelles and not to discontinue my operations. Your lordship will observe, that such a desperate action could not be fought, and such advantages could not be gained, without great loss; and I am sorry to add that ours have been immense.[13]

---

13   The Duke of Wellington, *Wellington's Official Despatch* (Waterloo), in, *The Book of War*, edited by John Keegan (New York: Penguin Books, 2000), pp. 178-84, citation p. 182.

In addition to the atypical risks attendant to military leadership, there is also a lifestyle that privileges camaraderie and group cohesion—as well as a certain degree of separation from the civilian world—as opposed to individualism and competitive Darwinism on the job. But more than this, what applies to the armed forces in developed Western democracies, at least, also applies to other professions like police that are specially authorized to use force on behalf of policy. We want honorable persons to staff the empowered coercive organs of the state because, absent a certain amount of self-regulation and public oversight, armies and police forces can be debauched into oppressive instruments of authoritarian rule.

Even when and where democratic control of the armed forces is established and secure, we want persons who have the physical power and legal authorization to kill to have consciences that will warn them off against murder, torture and other abuses of human rights. It is no coincidence that, during the George W. Bush administration, the persons who were most disturbed and vocal about allegations of torture and other indignities to prisoners in Iraq and other locations were American military lawyers. The dignity of humanity is an inseparable component of any military education. To see what happens when the preceding guidelines are violated, one need look no further than in those states where rising numbers of "child soldiers" are being recruited and trained to kill innocent civilians of another tribe or village. This is to sin twice: against humanity, and against humanity's most vulnerable members, whom armed forces should above all protect. When killing becomes wanton and random, armies have turned to mobs, and civilization into detritus. Distinctions between good and evil were as important to the ancient Greeks as they remain today, as Victor Davis Hanson reminds:

> While for the Greeks all wars presented only bad and worse choices, and were tragic in the sense of destroying the lives of young men who in peacetime had no intrinsic reason to murder one another, conflicts could still be judged as more or less good or evil depending on their causes, the nature of the fighting, and the ultimate costs and results. Some wars then were deemed better than others, and it was not all that difficult to make the necessary distinctions.[14]

Insights into civil-military relations also lead us into the recognition that, not only military or political leaders and their war aims, but also constitutions and polities, partake of moral excellence or failure. A highly-skilled and professional armed force, trapped within the framework of a nihilistic state and ideology, will march to ruin along with its suborned leadership. The German General Staff assumed that Hitler, once having been elected chancellor, could be domesticated by the prestige of the German Wehrmacht, the responsibilities of office, and the influence of centrist and pro-business politicians. This optimism was soon disappointed, in the fire of Hitler's demonically-inspired global ambitions and morally bankrupt

---

14   Victor Davis Hanson, *The Father of Us All: War and History, Ancient and Modern* (New York: Bloomsbury Press, 2010), p. 33.

definition of war aims. It is to their credit that a number of German officers chose to risk ignoring Hitler's orders for massacre or worse at the potential cost of their own lives. A smaller number resorted unsuccessfully to tyrannicide in their desperation to distinguish their professional ethos and personal moral compasses from those of the regime.

Democracies offer no guaranty against abuses of civil-military relations by politicians or commanders. Checks and balances against military overthrow of constituted civilian authority have prevented coups in the U.S. and in many other developed democracies. The deficiencies of modern democracies with respect to civil-military relations are more subtle than the risk of attempted coup. The more complicated issue is whether the political class and the military leadership can be sufficiently accepting of one another's professional perspectives and collaborative in the pursuit of shared policy aims. Madeleine Albright, then U.S. Ambassador to the United Nations and later U.S. Secretary of State in the Clinton administration, once ruffled military feathers by posing the query during Cabinet discussions: what's the point of having U.S. armed forces of such outstanding competency if we are not going to use them (in controversial peacekeeping, peace operations and other unconventional conflicts)? General Colin Powell, then Chairman of the U.S. Joint Chiefs of Staff, reports that he almost had "an aneurysm" in response to this question.[15]

According to Peter D. Feaver and Christopher Gelpi, significant differences separate the opinions of American civilian from military elites, with respect to important questions of whether and how to use force and, in addition, with regard to the "casualty sensitivity" of the respondents. On the matter of whether and how to use force, elite military officers "are more inclined toward a realpolitik view of the use of force—willing to use force for traditional national security threats like defense of allies and of geostrategic access to vital markets but more hesitant about using force for humanitarian missions and the "less-than-vital-interest" scenarios of intervening in foreign civil wars that have dominated the global agenda in the past decade."[16] Compared to elite military officers, elite civilians who have never served in the armed forces "are somewhat more interventionist, embracing a wider range of missions for the military," and nonveteran civilian elites "are more willing to use force gradually or incrementally." Elite military officers, on the other hand, are more skeptical about gradual escalation and are more likely to prefer overwhelming or decisive force relative to the military objective.[17] Feaver and Gelpi emphasize that these and other differences between civilian and military opinions are neither new

---

15   Peter D. Feaver and Christopher Gelpi, *Choosing Your Battles: American Civil-Military Relations and the Use of Force* (Princeton, NJ: Princeton University Press, 2004), pp. 2-3 and also citing Colin Powell with Joseph E. Persico, *My American Journey* (New York: Ballantine Books, 1995), pp. 576-7.

16   Feaver and Gelpi, *Choosing Your Battles*, p. 184.

17   Ibid.

nor surprising and influenced U.S. decisions on the uses of military force from early in the nineteenth century into the final decade of the twentieth century.

In response to Dr. Albright, General Powell might have said: "U.S. armed forces exist for the use and threat of force on behalf of national policy. American troops are not a colonial constabulary for cabinet and expeditionary wars, but the sons and daughters of American taxpayers who prefer to see them used for just causes, supported by a clear strategy for prevailing at an acceptable cost, and with a reasonable expectation of popular and Congressional support." He might also have pointed to the "Weinberger doctrine" promulgated by President Ronald Reagan's Defense Secretary Caspar Weinberger (with Powell's assistance) during the 1980s, with respect to the criteria for deciding upon and evaluating U.S. military interventions. To this, he might have added: "Although reasonable people can disagree about whether the Weinberger doctrine is too restrictive for some contingencies, the Weinberger guidelines are a necessary reminder that war is a uniquely dangerous and important decision, not only for those who risk their lives in battle, but also for states and leaders who engage in it." As Machiavelli warned, you can start wars as you prefer, but you may not be able to end them as you wish.

Indeed, with respect to civil-military relations, and all else related to strategy and national security policy, U.S. and other major power militaries have in the past two decades found themselves in a vortex of responsibility for hybrid wars or complex contingencies that mix conventional and unconventional military operations with diplomacy, state and societal reconstruction, cultural adaptation, and inter-agency cooperation both public and private.[18] As if this menu were not sufficiently complicated, the "war on terror" requires granular cooperation between civil and military intelligence providers and users, as well as across the compartments between "domestic" and "foreign" intelligence collection and estimation. "Strategic intelligence" has reappeared in the U.S. lexicon, at least as a term of endearment, referring not to "strategy" as defined in prior discussion here but to topside appraisals and inter-agency coordination.[19] The U.S. Department of Homeland Security alone provides challenges for the integration of in-house civil and military departments and agencies. As part of the more complicated inter-agency world after 9-11, the Director

---

18   On the concept of hybrid wars, see Owens, *U.S.-Civil Military Relations after 9/11*, p. 118 and p. 186. For related discussions, see: John Arquilla, *Worst Enemy: The Reluctant Transformation of the American Military* (Chicago, IL: Ivan R. Dee, 2008), especially pp. 132-81; Isaiah Wilson III, *Thinking beyond War: Civil-Military Relations and Why America Fails to Win the Peace* (New York: Palgrave Macmillan, 2007), especially pp. 35-57; General Tony Zinni and Tony Koltz, *The Battle for Peace: A Frontline Vision of America's Power and Purpose* (New York: Palgrave Macmillan, 2006), especially pp. 67-127; and Bevin Alexander, *How Wars Are Won: The 13 Rules of War – From Ancient Greece to the War on Terror* (New York: Crown Publishers, 2002), especially pp. 1-22.

19   A superior treatment of these and other intelligence issues appears in Gregory F. Treverton, *Intelligence for an Age of Terror* (Cambridge: Cambridge University Press, 2009).

of National Intelligence (DNI) supervenes over some 16 government agencies and departments, presumably sharing information and assessments in good time.

Further on, the list of challenges for future U.S. and allied planners includes the disturbing recognition that "war," and strategy related to war, now occurs in at least five domains: land, sea, air, space and cyberspace. The most recent of these possible domains for warfare offers up an interesting challenge for students of civil-military relations and other close followers of national security policy. Is "cyberwar" really war, even if no kinetic weapons are employed? What is the chain of command for unleashing cyberwar, and against what targets is cyberwar appropriate? Does a unique cast of cyber-warriors have to be created by each arm of service, specially trained in service doctrine and traditions? What, exactly, constitutes deterrence with respect to cyber conflict, and how applicable is deterrence in this domain?[20] Finally, and perhaps most meaningful for students of strategy, as related to civil-military relations: what are victory and defeat in cyberwar? Who certifies that cyber-surrender on the part of an enemy cyber attacker has taken place? Will war termination in cyberspace becoming an oxymoron, with 24/7 probing on the part of state and non-state actors into the vitals of others' network security and information stability?

The answers to the immediately preceding questions may provide fodder for numerous policy papers and dissertations. But in the actual world of policy making, they will necessitate the creation of matrix management, within and across departmental and agency lines. Geeks and gargoyles in the armed forces will need to coexist within a command climate that allows each to develop his or her professional skill set. The brute force and eternal climate of warfare (danger, exertion, uncertainty and chance, per Clausewitz) cannot be transcended, but only included in a broader appreciation of strategy and the art of war that will demand civil-military centurions of unprecedented vision and commitment. How many visionary commanders and policy makers can a country create? The answer is: as many as possible, and as soon as feasible. Otherwise, the cognitive complexity of future war, from botnets to bivouacs, will leave behind an industrial age civil-military relations in Washington and elsewhere. John Keegan may be incorrect in his assertion that battle may have abolished itself, but John Arquilla may be closer to the mark when he argues that the Pentagon is already past obsolescence.[21]

Plan of the Book

In Chapter 1, Isaiah Wilson III, Edward Cox, Kent W. Park, and Rachel M. Sondheimer note that three generations of current military leaders (co-)exist at

20   Expert discussion of this topic appears in Martin C. Libicki, *Cyberdeterrence and Cyberwar* (Santa Monica, CA: RAND Corporation Project Air Force, 2009).

21   John Keegan, *The Face of Battle* (New York: Penguin Books, 1976), pp. 331-43 on the abolition of battle. On the obsolescence of the Pentagon, see Arquilla, *Worst Enemy*, pp. 227-8.

any given moment, bringing with them different formative experiences and views on professionalism. In this chapter, they explore how generational differences help and hamper the transmission and evolution of contemporary understandings of the military as a profession within the context of civil-military relations. They examine three interwar periods (post-World War I, post-Vietnam, and post-Gulf War) to understand the importance of teaching, learning, and mentorship in overcoming potential gaps developed due to each generation's unique socialization into the military and society. Wilson, Park, Cox and Sondheimer contend that fostering dialog concerning professionalism within the military is as important, if not more so, to civil-military relations as engaging in similar conversations with civilian counterparts.

As Dale R. Herspring notes in Chapter 2, scholars working on civil-military relations focus excessively on the issue of control: to what degree do civil authorities control the activities of those in uniform? In fact, he argues, in established, stable political systems, the issue is not control: it is how to manage civil-military relations so that the relationship is one of "shared responsibility." Shared responsibility assumes the existence of conflict. Conflict is inevitable and good so long as it is controlled, and the military recognize the principle of civilian supremacy. According to Herspring, such a situation also assumes that civilian authorities respect the principles of military culture. In these circumstances the military feels free to express its opinion privately on issues related to national security. Conversely, if civilian authorities fail to respect military culture, conflict between military and civilian authorities will be exacerbated with a decrease in military willingness to provide and/or participate in national security decision-making. In this chapter, Herspring his comparative approach by focusing on civil-military relations in two countries: Canada and Germany.

What happens to military professionalism and civil polity in cases where authoritarian or other undemocratic regimes rule? In Chapter 3, Stephen J. Blank considers the case of post-Soviet Russian civil-military relations. The contemporary situation in Russia, according to Blank, presents a unique model of undemocratic control that starkly illuminates the dangers inherent in that form of civil dominance over the armed forces. The widespread corruption of the military, the politicization of multiple military forces to check each other and to increase the means of repression at home, along with the inherent tendency towards military adventurism visible in Russian policy, suggest that this form of control presents a clear and present danger to both Russia and her interlocutors, not to mention her neighbors. Blank's findings in this chapter have sobering implications for the relationship between Russia and the United States as well as NATO, including efforts on the part of U.S. and NATO foreign offices to "reset" security relations in a more collaborative, as opposed to a more confrontational, direction.

In Chapter 4, John Allen Williams argues that the international system is well into an era of conflict where the traditional notions of the "American Way of War" (clear enemy, moral clarity, conventional military focus) are decreasingly relevant. According to Williams, the United States military is especially aware of this, having sustained significant casualties in at least two wars still underway (including Iraq, where conflict will surely continue beyond the withdrawal of U.S. "combat" forces). From a civil-military relations perspective this period is notable by a partial yet massive mobilization that involves the active and reserve military components, but not the civilian society except insofar as reserve forces are now heavily involved. We have an "all recruited" military that does not reflect society in important ways, most notably the near-absence of societal elites in its ranks. Williams explores these phenomena and discuss their implications for the U.S. armed forces and for American civil-military relations in the future.

In Chapter 5, Damon Coletta considers the relationship between preferred strategies for research on U.S. civil-military relations and their ability to explain future success or failure in military adaptation to new global challenges and missions. According to Coletta, civil-military relations and their effect on strategy and policy are usually investigated with an emphasis on interaction at the highest levels of decision-making. Sociologists tend to approach this *problematique* from the bottom up while political scientists come from the top down, devoting attention to institutional structures and leadership style rather than attitudinal gaps or political activity in the society at large. In either case, changes in the high-level decision process are the mechanism by which civil-military relations affect strategy, policy, and eventually reform of institutional structures. Coletta notes that present military missions for the United States are undergoing profound change at low-intensity operations—counterinsurgency—and those potentially leading to the highest level of destruction—custody and maintenance of nuclear weapons. New operational tactics and coalitions plus new concerns from the national command authority are shifting the pressure points of civil-military relations as well as the kind of key cases emerging in the study of this crucial interaction.

Since the attacks of September 11, 2001, the United States government has made countering Islamic terror groups the central focus of its military and foreign policy efforts. Nevertheless, according to C. Dale Walton in Chapter 6, after nearly a decade—and despite the release of a great amount of written policy material, including several iterations of the National Security Strategy—Washington still does not have a coherent strategy that lays out a clear, realistic, and tangible long-term plan for countering terrorism and Islamist ideology more generally. In part, this fact simply reflects the difficulty of the challenges that Washington faces in countering Islamist terrorism. This also, however, is the result of a defective relationship between the civilian and uniformed components of the policymaking establishment. Walton argues that, in essence, the Executive Branch has failed, in both the Bush and Obama Administrations, to craft a "big picture" vision in which counterterrorism plays a role proportional to its importance in the overall grand strategy of the United States. The military, in turn, has remained focused on operational and tactical issues and has failed to press the Executive Branch for a clearer vision of the conflict formerly known as the Global War on Terrorism. The largely is a reflection of the U.S. military's traditional, and understandable, reluctance to involve itself in political controversies. The result, however, has been deeply unsatisfying from a strategic perspective: the vast expenditure of human and material resources has not necessarily resulted in greater security for the United States and "victory"—or even the criteria for defining victory in its unfocused counter-Islamist effort—remains elusive.

In Chapter 7, expert civilian and military analysts focus on the character of the U.S. armed forces and their relationships to the larger political system, society and culture. Gary Schaub, Jr. and Faculty Researcher and Defense Analyst Adam Lowther examine the relationship between the current U.S. All-Volunteer Force (AVF) (2010) and society by posing and addressing a series of tough questions. Who serves in the military? When the United States ended conscription and began to acquire its personnel voluntarily, significant concerns were voiced. Would the military attract sufficient and appropriate personnel? Would the self-selected force reflect American society in terms of demographics, socio-economic origin, personality, and ideology? Or would the force become increasingly separate and alienated from American society, maneuver to become politically independent from civil authority, and perhaps endanger the polity? Schaub and Lowther address these issues by examining the demographic makeup of the current U.S. military, including indicators of ideology and personality. They find that, although the U.S. military is indeed a unique institution in American society, its differences have not indicated the separation and alienation feared by many.

Armed forces require not only personnel and equipment of superior quality, but also a collective psyche of excellence manifest in their organizational DNA, including prevailing ideas about national security policy, grand and military strategy, and the conduct of military operations. According to expert analysts and scholars Jacob W. Kipp and Lester W. Grau in Chapter 8, the U.S. will be challenged in the present century to deal with security challenges for which no existing template suffices. The advent of network-centric warfare, and U.S. overall preeminence in information technologies, created the opportunity for rapid and decisive victory against regular armies such as those deployed by Iraq in 1991 and 2003. On the other hand, some of these same technologies, combined with unorthodox strategic thinking and operational-tactical practice, have enabled hybrid wars that blend conventional and unconventional motifs. Hezbollah's adaptive combination of regular and irregular warfare with social services and legitimate political activism offers an example of a highly capable fighting network. Therefore counterinsurgency theory and practice must contend with plural communities found within civilian populations that form the center of gravity for military strategy and operations. According to Kipp and Grau, for eight years the U.S. and its allies were committed to the conflict in Afghanistan without the benefit of a comprehensive strategy. Strategy is the responsibility of governments, not the military, but in too many instances the U.S. and other governments have abandoned their responsibility and left strategy to the generals, resulting in military- and geographic-specific strategies.

One noteworthy current, and past, societal impact on the U.S. military has been the use of business management methods, not only for the improvement of administrative efficiency, but also as conceptual templates for the art of war. In Chapter 9, Milan Vego describes, analyzes and critiques the practice of adopting various business models for some important and emerging U.S. warfighting concepts. Specifically, Vego examines (1) the Wal-Mart network and network-centric warfare (NCW)/network-centric operations (NCO) and its offshoot effects-based approach to operations (EBAO), (2) "just in time" and "sense and respond" logistics—and how it was applied in Afghanistan and Iraq, (3) emphasis on efficiency vs. effectiveness in force planning (especially in the U.S. Navy), and (4) the use of various business metrics in evaluating the post-hostilities/counterinsurgency phase of campaigns in Afghanistan and Iraq. In so doing, the author highlights some of the more optimistic DOD expectations for the transferability of business management into force planning and execution, including the potential for disconnects among ends, ways and means.

In Chapter 10, Stephen J. Cimbala examines the relationship between nuclear crisis management and information warfare or "cyberwar." Nuclear crisis management was a learned behavior for American and Soviet military and political leaders during the Cold War, with mixed results. There developed a shared recognition of the impossibility of prevailing in a nuclear war as between the two superpowers and their allies, and, therefore, in the priority of avoiding such a conflict. On the other hand, various studies of Cold War nuclear crisis management, especially those based on the Cuban missile crisis of 1962, revealed lapses in decision-making on both sides based on misperceptions, mistaken assumptions, uncertainty, "friction" à la Clausewitz, and other factors. The post-Internet world has enabled U.S. and other militaries with the capability for conducting extensive information operations, including cyberwar in support of conventional warfare or cyberdeterrence apart from war. However, the credible threat or actual carrying out of cyberwar during a nuclear crisis might exacerbate misperceptions and accelerate first-strike fears on the part of target states. Crisis managers and military operations fearful of information blackout could fell increased stress and uncertainty leading to a mistaken decision for nuclear preemption.

The concluding chapter summarizes the findings of the contributors to this volume and offers additional perspective on civil-military relations in the U.S. and more broadly. The lingering dilemmas of civil-military relations post-9-11 and post-Obama's first term are in some ways new, but in other ways classic. Policy makers and warriors are, as the French say, "condemned to succeed" together, but under technological and political circumstances in the twenty-first century that will differ in detail, if not in essence, from the past. The essence of viable civil-military relations is the production of good strategic effect, within the constitutional framework of the political system for which politicians and soldiers are performing their respective arts. Strategic effect is an abstraction, crossing the boundaries between policy and military operations for the threat or use of force. For example, lost wars tend to have deleterious effects on civil-military relations, although not necessarily destructive ones. Strategic effect is concrete as well as abstract; particular, as well as general, in configuration. U.S. strategic effect in counterinsurgency, as an example, differs in its particulars from effects sought in major conventional conflicts or in nuclear deterrence. In addition, in each case of strategic failure or competent performance, the "on the ground" specificity matters: against whom, where, for what reason, and with what constraints is the United States going to war, or practicing deterrence? In this regard, U.S. deterrence-defense practices and geostrategic aims overlap with and help to determine the outcomes for civil-military relations.

Chapter 1

Kids These Days: Growing Military Professionalism across Generations

Isaiah Wilson III, Edward Cox, Kent W. Park and
Rachel M. Sondheimer

The Army's expert knowledge can be broadly categorized into four capacities: Military-Technical, Moral-Ethical, Political-Cultural, and Human Development. Of the four, it is the human development capacity that sets the Army apart as a profession. As officers enter, develop, lead, and eventually retire, they have a profound impact on the institution as a cohort due to generational influences on organization and leadership. This chapter examines how generational differences help and hamper the human development capacity that the Army must have to socialize, train, educate, and develop the Army officer corps to be stewards of the profession.

Three generations of current Army leaders coexist at any given moment, bringing with them different formative experiences and views on professionalism. The procession of these three groups of people will profoundly shape the operation and legacy of the institution long after their respective tenures. The manner in which each group of leaders shapes the Army will have much to do with their own formative experiences rising through the ranks. In the halls of the Pentagon today, these generations take on labels such as, "Gulf War Generals, Bosnia/ Kosovo Colonels, and Iraq/Afghanistan Captains and Majors." Each of these groups corresponds to a larger societal generational group—Gulf War Generals are predominantly Baby Boomers, Bosnia/Kosovo Colonels are members of Generation X, and Iraq/Afghanistan Captains and Majors are Millennials. A closer look at these three populations reveals much about the formidable experiences that shaped their professional view:

> Boomers: Born between 1946 and 1964, this group of around 77.3 million individuals came of age during a period of significant social and political transition.[1] The generation itself straddles two distinctly different periods: the 1950s, when society was still deeply-rooted in traditional values of stability and responsibility, and the 1960s and 1970s, a time of significant

---

1   U.S. Census Bureau: Baby Boom Population, USA and by State (July 1, 2008), http://www.boomerslife.org/baby_boom_population_us_census_bureau_by_state.htm (accessed October 29, 2010).

social and political turmoil in our society. From the Civil Rights Movement to the Vietnam War, this generation witnessed and experienced the effects of the rebellious counterculture lashing back at authority. Within the officer corps, the Boomers make up most of the senior general officers with the youngest of this generation reaching 30 years of service by 2012. While the oldest members of this cohort were commissioned during the Vietnam era, most of Boomers' careers as officers started in the 1980s at the beginning of the Reagan administration's new military build-up. They experienced the post-Vietnam professionalization of the Army with large investments in new technology and equipment. As Lieutenants and Captains, they trained and prepared for the Soviet invasion through the Fulda Gap only to see their adversary collapse without a shot fired. Instead of the Soviet armored columns, this generation of officers fought in the desert against Saddam Hussein during the Persian Gulf War as senior Captains and Majors. Their careers continued as Lieutenant Colonels and Colonels with some of the older cohorts making general officer during Somalia and Kosovo. Most who continued in active duty service were general officers when 9/11 occurred.

Generation X—Born between 1965 and 1980, this group of 46 million individuals is sometimes known as the MTV generation.[2] While the Boomer generation came of age during dramatic social change, Generation X came of age during a dramatic technological change. New innovations in technology, such as faxes, copiers, and computers, fundamentally changed the way people lived and worked. Within the officer corps, Generation X currently makes up most of the field grade officers with some of the older cohorts starting to become general officers. Mostly commissioned after the Cold War, the Persian Gulf War was the first testing ground for some of the older cohorts while "Military Operations Other Than War" became the norm, somewhat reluctantly, for the younger cohorts. Unlike the Boomers and other generations, this population of officers did not share a common experience of war in the traditional sense of having a common adversary. While experiencing an increase in operational tempo, they were engaged in a variety of peacekeeping, peace enforcement, and humanitarian missions. This changed after 9/11 when this generation of officers provided the bulk of tactical leaders in Afghanistan and Iraq. Almost all had served multiple combat tours by the time they reached the rank of field grade officer.

Millennials: Also known as Echo Boomers, Generation Y, and Generation Next, this group of individuals was born between 1980 and 1994. Most are just beginning to enter the work force. At approximately 76 million, they constitute one of the largest generations since the Greatest Generation of World War II.[3] Whereas the previous two generations were digital immigrants

---

2   U.S. Census Bureau: http://www.census.gov (accessed October 29, 2010).
3   Ibid.

who had to learn and adapt in the information age, the Millennials are digital natives. They do not remember a time without computers, the Internet, cable TV, and cell phones. For the Millennials, multitasking is the norm and they feel perfectly comfortable simultaneously watching YouTube, reading an email, chatting on instant messenger, and updating a Facebook status, all while listening to music on an iPod. Most do not remember a world before 9/11, when people did not have to take their shoes off before boarding a plane. Most Millennials joined the Army at War and have little concept of a peace-time Army. Making up almost the entire population of Lieutenants and Captains, Millennials bore the brunt of the tactical fight in Iraq and Afghanistan. They do not understand when older generation officers talk about a "normal" rotation through the national training centers. For the Millennials, counterinsurgency and counterterrorism is the norm. Millennial officers are highly tactically competent, battle-hardened, and confident in their ability to conduct operations independently of higher headquarter command and control. Because of this, they are understandably "irreverent" to hierarchical command and control. They are tactically talented as battlers but often immature in their understanding of and appreciation for the operational and strategic-level.

One difference between the Boomers and Millennials is highlighted above—*the degree of autonomy that each generation is comfortable with*. Boomers grew up in an Army where the platoon/company often moved with the brigade/division as a whole. Millennials are comfortable in working autonomously even from their own battalion; they see that as the norm. This is not limited to their military experience; Millennials have been educated collaboratively even as children and are more accustomed to collaboration than hierarchy.[4]

One's generational perspective profoundly influences future decision-making and leadership style. The promotion from company grade officer to field grade officer is one of the more difficult transitions one must make during an Army officer's professional career. Some never quite make the transition and continue to operate with perspectives stuck at the tactical level. The Army's promotion and command selection system reinforces this behavior by (over)relying on tactical performance as key indicators for strategic potential. It should not be surprising, then, that field grade officers look back and rely on their tactical experiences, consciously or subconsciously, to help them analyze new situations. It is this world view, formed early in the career progression, that provides professional perspective on different courses of action. As such, while it is difficult and, in some cases counter-productive, to label individual officers based on their generational background, understanding the formative milestones for these different populations

---

4   Rhonda Smillie, *Suitability of Millennials to Lead the Profession of Arms* (Carlisle Barracks, PA: U.S. Army War College, 2010), p. 17.

can help us better understand aggregate behavior and interactions among the various levels of the officer corps.[5]

In the face of the coexistence of these three vastly different generations under the aegis of the "current Army leadership," how do we communicate and develop a single contemporary professional ethos? As an organization, the Army must maximize the transmission of each cohort's expertise among the other generations. For example, the senior leadership brings years of experience that it must relay in a top-down fashion to the younger cohorts while the junior leadership brings knowledge of the current fighting force that is of use to its superiors. How is this knowledge best communicated as a means of shaping the current and future Army profession?

The Importance of Teaching, Learning, and Mentorship

Dialog and discourse among the generations are the keys to shaping a cohesive professional ethos within the Army. Generally speaking, institutions must allow for generations to teach and learn from each other in formal and informal settings. Moreover, this teaching and learning must occur from the top down, the bottom up, and from peer-to-peer.

These relationships and communication styles must take on a mentorship as opposed to coaching model. Coaching involves the passing of knowledge from previous generations to the next under the assumption of a stagnant environment in which there exists a known and finite answer that can be imparted to the next generation. Such coaching is usually undertaken by those no longer in the profession. In contrast, mentorship involves the distillation of an approach to incorporating knowledge and cultivating a way of thinking as one adapts to a changing environment. Here, there is no known or finite answer, but there is a right way to think about problem-solving and the cultivation of ethics to shape behavior. Such mentorship is usually undertaken by active but senior players in the profession.

Case Studies of Interwar Periods

To emphasize the importance of mentorship and dialog across and within coexisting generation, we present short examinations of the key advances in the cultivation of Army professionalism during three interwar periods. Interwar

---

5 For a detailed analysis of Boomers and Generation X, see Leonard Wong's monograph *Generations Apart: Xers and Boomers in the Officer Corps* (Carlisle Barracks, PA: Strategic Studies Institute, 2000). For a detailed analysis of Millennials, see Rhonda Smillie's monograph *Suitability of Millennials to Lead the Profession of Arms* (Carlisle Barracks, PA: U.S. Army War College, 2010).

periods allow for time for self-reflection and collection of lessons learned from the most recent conflict. Interestingly, leaders cannot obtain an adequate assessment of these lessons unless there is communication between and among the different generations of officers—fighting forces on the battlefield, mid-level officers commanding on the ground, and key leaders strategizing from a certain distance. These vignettes highlight what we can learn about the importance of teaching, mentorship, and dialog in the cultivation of the professional ethos from each of these formative periods.

Post-World War I to World War II

Budget cuts made the Army a hollow shell throughout the 1920s and 1930s. The National Defense Act of 1920 authorized a force of 18,000 officers and 280,000 men, but the actual strength of the Army was less than half this number. It was common for a rifle company to have only seven or eight men available for duty. In 1932 the chief of staff, Douglas MacArthur, reported that both Belgium and Portugal had larger armies than the United States.[6] Forced to do more with less, the officer corps renewed its focus on professionalism, building on the reforms of Secretary Elihu Root in the days following the Spanish-American War. Mentorship from above played a key role in officer development. Junior and mid-level officers, many of whom were veterans of the recent conflict, were encouraged to research and publish articles in military journals, which flourished during this time. In two famous examples, both George Patton and Dwight D. Eisenhower were encouraged by Brigadier General Fox Conner to publish articles in the *Infantry Journal* in 1920.[7]

The War Plans Division of the General Staff undertook a review of the Army's officer education system, based on input from Newton Baker, the Secretary of War. Reflecting on the American experience in World War I, Secretary Baker wanted officers for the General Staff who possessed a "broader knowledge, not only of their purely military duties, but also a full comprehension of all agencies, governmental as well as industrial, necessarily involved in a nation at war."[8] At every level, officers were encouraged to question basic assumptions and develop critical thinking skills through the Army's educational institutions. During this time, at the United States Military Academy, under the leadership of Herman Beukema, Professor of Economics, Government and History, cadets began to study international relations for the first time using a comparative methodology.[9] The

---

6   Edward M. Coffman, *The Regulars: The American Army 1898-1941* (Cambridge, MA: Harvard University Press, 2004), pp. 233-4.

7   Carlo D'Este, *Patton: A Genius for War* (New York: Harper Perennial, 1995), p. 297.

8   Harry P. Ball, *Of Responsible Command* (Carlisle Barracks, PA: U.S. Army War College, 1983), p. 151.

9   Theodore J. Crackel, *West Point: A Bicentennial History* (Lawrence, KS: University Press of Kansas, 2002), pp. 202-3.

Army War College was separated from the General Staff and two schools for junior officers were re-established at Fort Leavenworth. All three schools emphasized the need for effective staff planning to collaboratively solve a hypothetical military problem, culminating in a war game exercise. Not all officers were prepared for such a curriculum. Of the 78 officers in the Army War College class of 1920, 10 did not complete the course and did not receive credit for their attendance. Three others completed the course but were not recommended for either command or duties on the General Staff.[10]

During this interwar period, budget constraints and the organization of the Army's institutions provided a space for the different generations in the officer corps to teach and learn from each other in both formal and informal settings. The mentorship approach, which is distinctly different from a coaching communication style, facilitated and reinforced bonds of camaraderie and trust that would establish a cadre of professional officers as World War II began.

Post-Vietnam through the Gulf War

The period immediately following the Vietnam War was a tumultuous time not only for the U.S. Army but the entire nation. Racial tension, rampant drug use, growing disillusionment of the political system following high profile assassinations and political scandals all served to undermine the institutional foundation of our society. It was during this turbulent and chaotic time that the Army shifted to an all-volunteer force (AVF). This began a series of reforms within the U.S. Army that significantly altered the future of the force and necessitated a reliance on mentorship and education of its ranks.

An emerging trend resulting from the end of the draft on July 1, 1973 was the increasing reliance on women to fill the ranks of the AVF.[11] The initial recruits in the AVF failed to meet expectations in quality and quantity with record number of category IV recruits, the lowest category of enlistment on the Armed Forces Qualification Test. Integrating women into the ranks brought in highly qualified recruits, most with high school diplomas, to make up for the shortages in qualified male recruits.[12]

Despite the best efforts of the Army, the 1970s became known as the lost decade. An internal report by BDM Corporation for the Pentagon stated in 1973, the Army was "close to losing its pride, heart, and soul and therefore [its] combat effectiveness."[13] In 1979, General Shy Meyer, Chief of Staff of the Army informed

---

10   George S. Pappas, *Prudens Futuri: The U.S. Army War College 1901-1967* (Carlisle Barracks, PA: U.S. Army War College, 1967), p. 104.

11   Bettie J. Morden, *Women's Army Corps, 1945-1978*, U.S. Government Printing Office, First Edition (November 1990).

12   Ibid.

13   The BDM Corporation, *A Study of Strategic Lessons Learned in Vietnam: Omnibus Executive Summary* (McLean, VA: BDM Corporation, April 28, 1980) EX-8.

President Carter, "Mr. President, basically what we have is a hollow Army," as he reported that he had neither the divisions nor the lift capability to reinforce U.S. forces in Europe with 10 divisions in case of a Soviet attack.[14] Only four of the 10 active divisions in the U.S. were capable of deploying overseas in an emergency, and the force was plagued by chronic drug and alcohol abuse as the number of recruits with a high school diploma fell to its lowest point since transitioning away from the draft.[15]

The impact of this stress on the force in this transition period opened lines of communication between mid-level officers and their superiors. With their recent combat experiences fresh in their minds, mid-career officers became increasingly vocal in expressing their dissatisfaction with senior Army leaders and the bureaucracy. Some of this feedback made its way to a select number of senior officers who saw the need for extensive reforms and were willing to listen to the suggestions of their subordinates. One such officer was General William DePuy who oversaw a drastic reorganization of the Army in which the Continental Army Command (CONARC) was divided into Forces Command (FORSCOM) and Training and Doctrine Command (TRADOC). Breaking TRADOC away into a more independent center for learning and development allowed it to flourish. New doctrine and radical new ideas on training emerged including the development of National Training Centers that incorporated realistic war games using high tech training aids like MILES (Multiple Integrated Laser Engagement System). This was a drastic departure from the traditional training model of ranges and classroom instructions.

Leaders also reacted to changes in the Army by creating new loci for study and reflection and by trying to reshape the identity of the youngest members of the force. Key leaders, including General Walt Ulmer, who risked his career with a scathing rebuke on the Army with *Study on Military Professionalism*, were empowered by the Chief of Staff of the Army, General Meyers, to spearhead the effort to reinvigorate the study on leadership and professionalism. To boost the number of quality recruits joining the Army under the AVF, General (retired) Max Thurman better aligned recruiting strategies and tactics with the motivations and interests of younger generations with a new marketing message, "Be All You Can Be."[16] These leaders acted as champions for new and progressive ideas emerging within the ranks. They invested time and energy in listening and building upon the advice of their subordinates and, in some cases, risked their careers to shift the culture of the Army profession. Ultimately, they were successful in establishing a new framework from which to remake the Army and paved the way for younger generations.

---

14   James Kitfield, *Prodigal Soldiers* (New York: Simon & Schuster, 1995).

15   Ibid.

16   Ibid.

Post-Gulf War to 9/11

On February 28, 1991, coalition forces led by the U.S. defeated Saddam Hussein
and the world's fifth largest Army, just 100 hours after the start of the ground
invasion. In many ways, it was a validation of the strategic shift and the investments
made over the past two decades. Doctrine, training, equipment, personnel, and
leadership all came together to signify the rebirth of the U.S. Army from the
shadows of the Vietnam War. The stunning success reinforced the traditional view
of war as conventional threats requiring advanced technology and overwhelming
use of force. Development of unconventional capabilities to meet asymmetric
threats was largely marginalized even as the Army deployed on an increasing
number of *Military Operations Other Than War*.

The domestic political landscape in the immediate aftermath of the first Gulf
War was challenging and reflected the typical American postwar reaction—a
dramatic downsizing of the force in expectation of a cost-savings peace dividend
that could be applied to pressing domestic needs as the economy emerged from
recession. Indeed, given the overwhelming military success, America's leaders
and citizens considered the armed forces to be overly capable for the perceived
future security environment.

The absence of any clear recognizable threat during this period of time
encouraged the perception that it was prudent to reduce the armed forces. Thus,
budget constraints forced the military to balance its efforts between maintaining
readiness and fielding new capabilities to deal with the growing array of unknown,
but suspected, threats. These conditions compelled the Army to man, equip, and
train a military force capable of providing for the common defense, but "on the
cheap" and in a traditional mechanized force-design fashion.

During this interwar period, the Boomer generation discussed earlier in this
paper served as field grade officers and members of Generation X served as
platoon leaders and company commanders. Training, education, and mentoring
was robust, with most units conducting officer professional development (OPD)
and Non-Commissioned Officer Professional Development (NCOPD) sessions on
a regular basis. Almost all of this training, however, was within the context of
the success the Army enjoyed in Operation Desert Storm. As units increasingly
became involved in operations other than war (OOTW), the prevailing mentality
continued to view these operations as a sideshow to the main event, a major
regional war.

Era of Persistent Conflict

With the attacks perpetrated on September 11, 2001, the United States entered
what is now often characterized as an era of persistent conflict. In retaliation for
these attacks, the U.S. initiated major combat actions in Afghanistan and Iraq.
Each of these operations developed into counterinsurgency battles which are now
in their tenth year and continue to shape the U.S. Army's personnel, training, and

resource policies. The increasing challenges of non-state actors to the traditional Westphalian nation-state system, combined with the threat of cyber-warfare, globalization, competition for scarce resources, and the rise of emerging powers such as China, Brazil, and India, have significantly altered the Army's focus on conventional or traditional large-scale war in ways that dwarf the OOTW missions of the previous decade.

The predictable training rotations and professional development programs of the previous era have been replaced by multiple deployments to protect American interests abroad. It is during this period that the labels discussed at the beginning of this chapter apply. Millennials are leading platoons and companies conducting full-spectrum operations. These operations combine the offensive and defensive operations that are a distant memory for the Gulf War Generals with the stability and civil support operations that Generation X Colonels conducted in their youth. Furthermore, these experiences will be formative for the Millennials who will one day replace their superiors as the Army's strategic leaders. How the Army responds to the challenges of intergenerational communication will determine whether the resulting leadership is positive or negative.

The Road Ahead

A review of Army introspection during three key interwar periods highlights the necessity of education and intergenerational communication as the military reacts to an ever-changing landscape in this era of persistent conflict. Moreover, the vignettes emphasize the importance of focusing teaching, training, and mentorship on the internal dynamics of the institution, especially concerning the creation and maintenance of a professional organization.

The Army will enter another transformative interwar period as we approach the end of operations in Afghanistan. The generational gap in this period will be exacerbated by post-9/11 conditions of new enemies, new battle-spaces, and new kinds of wars. It will also be affected by the force redesigns of Army Transformation and the shift from the Army of Excellence (AOE) air-land-battle designs, premised on the Division as the basic warfighting unit, to the Modular Force, where plug-and-play is the operational and organizational foundation, and the BCT is the new baseline warfighting unit. It is clear from the case studies above that every generation of junior officers has a sense of disconnect from the older generation, a feeling that their elders "don't get it." Communication, education, and mentorship go a long way towards ameliorating this sense of disconnect. The generational gap, however, is more stark today than it has ever been. It is imperative that the Army creates a climate of communication across the three generations currently present in the Army to develop the officer corps that will lead the "next Army," leveraging the expertise and experiences of each of these cohorts.

As important as the method of dialog across and within the generations of leaders coexisting within the Army at any given moment is the substance of those discussions. As such, we conclude this article with six key topics and underlying questions that must inform contemporary and future consideration in the development of the professional Army officer:

- *The Soldier and the Policy Process*: What does it mean to be a military professional in the twenty-first century? How do we instill a notion of professionalism in the current and future officer corps? How can the military officer provide policy advice borne out of expertise while maintaining partisan neutrality and avoiding partisan policy advocacy?
- *The Soldier and the Military-Industrial-Congressional Complex*: Does the nature of military professionalism change in war versus peacetime and how does perpetual war affect this dynamic? What are the consequences on national security policy of either the obsolescence of military professionalism or eroding objective control?
- *The Soldier and the Strategy-Making Process*: How does the changing threat environment impact the strategy-making process? Does the military have the necessary jurisdiction, legitimacy, and the expertise to fulfill our professional obligation to our nation in respect to "new frontiers," for example, cyber security?
- *The Soldier and the Political Campaign*: What is the proper balance between the professional soldier and the active citizen as embodied by the citizen soldier? Should military professionals abstain from voting in elections determining their commander-in-chief? What are the effects of the contemporary coexistence of the perpetual campaign and the perpetual war?
- *The Soldier and the Military-Media Complex*: What is the role of the media in shaping perceptions of the military in the policy process and of military professionalism? What challenges do contemporary war and military coverage pose to the state-soldier relationship? How can we balance the media's natural inclination towards openness with the military's often necessary desire for the secrecy and security of information?
- *The Soldier and Society*: What are the effects of changing military demographics on the military's relationship with and integration into American society? How does the military adapt to changing social mores and how does this influence the military's role in the policy process and in society at large?

Continual dialog and debate among the three generations of leaders concerning the proper role and function of the professional military officer within these six areas will allow for the Army to adapt to a changing world while not losing its core mission and place within the republic.

Chapter 2

Searching for a More Viable Form of Civil-Military Relations: The Canadian and American Experiences

Dale R. Herspring

Specialists in civil-military relations are always looking for the "optimal" form of interaction between civilians and the uniformed military. Traditionally, such scholars have focused on political control. The idea is that unless strong controls are placed on the uniformed military, it will have an unhealthy impact on policy and may even seize control of it.

This chapter takes a different approach. It argues that the optimal form is one of "shared relationship" between senior civilian officials and senior military officers.[1] Such a relationship assumes a number of conditions: Senior military officers in a mature, institutionalized, stable polity, accept civilian control: it is a given. They are compliant and obedient. Second, military officers feel free to provide honest advice without fear of career or personal reprisals and are willing to do so. Third, senior civilian authorities respect and are interested in listening to what senior military officers have to say about national security affairs although civilians always have the final say on policy. The civilians are under no obligation to accept military advice.

A policy of "shared responsibility" also maintains that a conflictual relationship is not only normal, but it is positive and healthy, provided it is regulated. Eliminating conflict would remove the give and take that is a critical part a shared relationship. Conflict is ubiquitous. It is the engine that drives the national security decision-making process: provided the criteria noted above are present.

The biggest problem with a focus on control is that it may not only distort the process, it provides little insight into the nature of interactions between the military and its civilian masters. As Douglas Bland observed, "No coup, no problem, and so no further discussion is required."[2] The bottom line in this chapter is that civil-military relations is about more than coups; that is certainly true if one is looking at the issue in a comparative fashion. Political control means different things to different people.

---

1   The term "shared relationship" is taken from an article by Douglas Bland, "A Unified Theory of Civil-Military Relations," *Armed Forces and Society*, 26:1 (Fall 1999): 8.

2   Bland, ibid.

With the exception of American General Douglas MacArthur, who openly disobeyed orders from the highest military and civilian authorities during the Korean War, or Canadian Rear Admiral William Landymore, who openly defied Defense Minister Hellyer over service unification, neither the U.S. nor Canada has faced the problem of open military rebellion in recent years. There have been cases when officers from both countries have publicly objected to policy issues, but in almost all cases serving officers who dissented from policy were reprimanded. MacArthur was relieved of command by President Harry S. Truman much to the relief of senior American officers, and Landymore was forced into retirement. In short, in neither case has there been a threat of a coup.

Given the inevitability of conflict, the question arises, how best to manage it? One of the best-known suggestions comes from the late Samuel Huntington. In his seminal work, *The Soldier and the State*,[3] he argued that politicians should determine the country's policy and goals, while the military professionals implement their orders. A senior officer may advise, but he may *not* become involved in policy-making. It is the task of a military professional to learn the mechanics of war, to train his troops so they can carry out the missions assigned to them by the civilian leadership. Or, as he put it, the primary task of the professional soldier is "the direct operation, and control of human organization whose primary function is the application of violence."[4] The nature of war has become so complex and technical that it is necessary to have these specialists manning the armed forces. They are the only ones who know how much and when to apply force.

This writer disagrees with Huntington. Military influence is *not* the critical factor in the civil-military relationship. There is no doubt that military influence is important—that is, to what degree is a general able to persuade his civilian counterpart to purchase this weapon or engage in the use of military force? More important is the nature of interaction between civilian and military leaders that helps define "healthy conflict."[5]

In most cases, political-military decision-making is not a zero sum game. Instead, as Gibson and Snider noted, when it comes to most issues, the two sides will be involved in what they called, an "Area of Overlap and Tension." As they stated with regard to the American case, "To our understanding, such decisions are most often made (particularly within the multilevel "interagency") in a more informal, collegial manner during which the positions of both senior civilian and military leaders are fully vetted and debated."[6] There are times

---

3   Samuel Huntington, *The Soldier and the State* (Cambridge: Cambridge University Press, 1957).

4   Ibid., p. 11.

5   For an essay arguing the critical importance of "influence," see Kobi Michael, "The Dilemma Behind the Classical Dilemma of Civil-Military Relations," *Armed Forces and Society*, 33:4: 518-46.

6   Christopher P. Gibson and Don M. Snider, "Civil-Military Relations and the Potential to Influence: A Look at National Security Decision-making Process," *Armed*

when the military officer is wrong, and vice versa. Having both sides respect the other, and having the military willing to point out when that officer sees problems with the civilian's thought-process makes for better decisions. Or as Bland put it, in discussing a "shared responsibility,"

> Civil control of the military is managed and maintained through the sharing of responsibility for control between civilian leaders and military officers. Specifically, civil authorities are responsible and accountable for some aspects of control and military leaders are responsible and accountable for others. Although some responsibilities for control may merge, they are not fused. The relationship and arrangement of responsibilities are conditioned by a national evolved regime of "principles, norms, rules and decisionmaking procedures around which actor expectations converge" matters of civil-military relations.[7]

Thus, instead of always looking at matters from the perspective of the civilian leadership in its efforts to "control" the military, I will look at the process from the opposite point of view: that of the generals and admirals. As noted above, this is not to suggest that they reign supreme. It maintains that by focusing on how the generals and admirals react to the decision-making process, one may be better able to understand that process.

From a theoretical standpoint, it is important to emphasize that Bland's shared responsibility is an ideal type. It is doubtful we will ever encounter a situation where the shared responsibility works perfectly—where there are few if any areas of disagreement between civilians and soldiers, but the goal is to create a situation where the relationship is symbiotic. As Sarkesian and Connor commented with regard to the American experience, "The problem is to develop a relationship that is appropriate and acceptable to both civilians and the military, while insuring that the military has an appropriate and realistic role in the decision-making process."[8]

Military Culture

This raises the question, how does one know if civil-military relations are a "shared responsibility" from the perspective of the military? The answer is how civilians deal with military culture. For example, do they respect it or intentionally ignore it?

To begin with, it is important to note that military culture is different from civilian culture. In the military, members are socialized so that they learn how to "act properly" inside the confines of the various services. To a large degree,

---

Forces and Society, 25:2: 195.

   7   Bland, "A Unified Theory of Civil-Military Relations."

   8   Sam C. Sarkesian and Robert E. Connor, *The US Military Profession into the Twenty-First Century: War, Peace, and Politics* (London: Routledge, 2006), p. 19.

military culture is a result of the organization's effort to prepare its members for its end goal. Inside the military, the organization's mission is often referred to as "to kill and break things." This means that the military's organizational culture, by definition, is structured to enable the military to fight wars. In the process of fighting a war, military service is unique because in carrying out his or her mission, the soldier must act in a certain manner and may be required to lay down his or her life. In this sense, military culture is different from culture in other organizations.[9]

Military culture is critical to well-functioning armed forces. As Sarkesian and Connor put it, "The military profession stands and falls according to its ability to maintain and reinforce military culture."[10] In an abstract sense, military culture refers to things like honor, devotion to duty, service to the nation, and subordinating oneself to those in command—all the way up to the senior official in charge, be he president or prime minister.

Operationalizing Military Culture

Military Culture is the key to understanding the concept of "shared responsibility." As noted previously, there will always be a conflict between the norms of military culture and those of the civilian leadership. However, if the norms of military culture *are not respected*, military leaders will be less likely to speak openly and provide their honest views on critical issues to the political leadership. If, on the other hand, the norms of military culture *are respected*, the level of conflict may decrease because the officers will believe they have had their day in court.

The following indicators of military culture are based on a wide variety of U.S. and Canadian sources.[11]

---

9   One may argue that para-military forces such as the police also require the individual to put his or her life on the line in carrying out his or her duties. That is tree, and that is why some of the military's cultural characteristics are shared by police forces. However, the military is more removed from the civilian sector and is trained to use the more complex forms of violence. For a discussion of civilian military culture, see Thomas Langston, "The Civilian Side of Military Culture, *Parameters* (Fall 2000): 21.

10   Sarkesian and Connor, *US Military Profession into the Twenty-First Century*, p. 79.

11   For Canada, see Allan D. English, *Understanding Military Culture: A Canadian Perspective* (Montreal: McGill-Queen's University); M.D. Chapstick, "Defining the Culture: The Canadian Army in the 21st Century," *Canadian Military Journal* (Spring, 2003); David Bercuson, *Significant Incident: Canada's Army, the Airborne, and the Murder in Somalia* (Toronto: McClelland and Steward, 1996); D.C. Loomis and D.T. Lightburn, "Taking into Account the Distinctness of the Military from the Mainstream of Society," *Canadian Defence Quarterly*, 10 (1980). For the U.S., see A.J. Bacevich, "Tradition Abandoned: America's Military in a New Era," *National Interest*, 58 (Summer 1997), 3; John Allen Williams, "The Military and Modern Society," *The World and I* (Summer 1999), 311; The Military Edgar R. Puryear, *American Generalship: Character is Everything*:

1. Executive leadership: All militaries are hierarchical. The private looks to the sergeant for direction, the sergeant to the captain, the captain to the colonel, and the colonel to the general. Equally important, the general, especially at the highest level, looks to the civilian leadership for leadership. It should be emphasized that not only do military officers accept civilian leadership, they expect it. A failure to provide such leadership will leave the senior military leadership confused and more than not when that happens there will be confusion inside the armed forces.

2. Respect for the military: Many years ago, Richard Betts commented that military leaders are expected to play a role in civil-military decision-making. How much of a role they play depends on the president, prime minister, secretary of state or defense minister. What is important, he noted, is that when they are ignored they become "alienated from their administrative superiors."[12] Furthermore, as he noted, "Military leaders become alienated from their administrative superiors in direct proportion to the decline in their direct influence and their perception of the gap between their rightful and actual authority."[13] This does not mean they are about to launch a coup or that they will shirk when given an order. A civilian leader can issue an order and it will be carried out, even by the most unhappy officer—that is part of their job. However, if civilian officials fail to respect them, their willingness to offer advice will inevitably be impacted negatively. The final decision is with the civilians, but ignoring the generals and admirals will only alienate them and undermine chances for a shared responsibility.

3. Clear orders and chain of command: Because the military is so hierarchical, officers expect civilians to obey the chain of command. That means that the civilian's orders are clear—they avoid the obfuscation that seems endemic to many civilian institutions—and they are issued by the most senior authority downward through officers based on rank and position.—from the most senior to the most junior. Too often, military officers, especially in the U.S.—have complained about coming out of a meeting with the president, totally confused on what it was that he expected them to do.

The Art of Command (Novato, CA: Presidio, 2002), pp. 1-43; Richard K. Betts, *Soldiers, Statesman and Cold War Crisis* (New York: Columbia University Press, 1977); Peter Maslowski, "Army Values and American Values," *Military Review*, 70:4 (April 1990): 10-23; Richard H. Kohn, "How Democracies Control the Military," *Journal of Democracy*, 4:8 (1997): 140; Thomas E. Ricks, "The Widening Gap Between the Military and Society," *Atlantic Monthly* (July 1997), 66-67; "The Cultural Demolition in the Military, *Washington Times*, November 20, 1998; Peter Feaver, "The Gap, Soldiers, Civilians and the Mutual Misunderstanding," *National Interest*, 61 (Fall 2000); Gregory D. Foster, "Failed Expectations: The Crisis of Civil-Military Relations in America," *Brookings Review* (Fall 1997), 46-8; and Elliot Cohen, *Supreme Command: Soldiers, Statesmen, and Leadership in Wartimes* (New York: Free Press, 2002).

12   Betts, *Soldiers, Statesmen and Cold War Crisis*, p. 5.
13   Ibid., p. 8.

4. Even if the civilian executive does provide leadership, if its orders are unclear, and the senior military officers do not have the option of asking for a clarification (or if the executive is not available for such a clarification), there is a danger that the military may not act as the political authorities wished. Second, failure by civilian authorities to follow the chain of command may undermine authority and negatively impact a military operation.

5. Respect for the military nature of its organizational structure: The military believes that carrying out its unique tasks requires a special culture which includes its own set of symbols, traditions and customs. It will strongly resist efforts to "civilianize" it because it believes that these symbols and tradition are critical to its ability to do its job.

Canadian and American Security Concepts and Institutions

One of my colleagues suggested that given the institutional and historical differences between the U.S. and Canada, not to mention the difference in deployable military power, such a comparison was non-starter.[14] Furthermore, someone asked, why compare the two countries? In answering this question, the paper proceeds from the old dictum that "a person who knows only one country knows no country.[15]

Given these differences, which are indeed significant, how can the civil-military experience in the two countries be compared? The answer lies in looking at incidents in all four categories noted above: Executive Leadership, Respect for the Military, Clear Orders and Chain of Command, and Treatment of Symbols and Traditions. It should be noted that no effort is being made to draw any statistical correlation with this material. The findings will be based on the preponderance

---

14   1. Canada has a lack of traditional enemies. The U.S., on the other hand, has world-wide commitments and seems to never lack enemies. 2. Canada knows that if it is threatened by Russia or anyone else, U.S. national interest dictates that Washington will come to its defense. 3. The Defense Department is a powerful bureaucratic player in Washington. That is not the case in Ottawa. Defence secretaries seldom take the job seriously, while waiting for an opportunity to move to a more powerful position. 4. Since World War II, there has been a lack of public support for Canada's military involvement around the world, while in the U.S., support for the military as a whole has remained high. 5. Both countries have faced ethnic problems: the U.S., the integration of blacks into the U.S. military, while Canada faced the French "problem." American blacks were not anti-military as were a large number of French Canadians, and integrating them into the U.S. military has been much easier and smoother than for French Canadians into CF. 6. While the U.S.'s security policy may appear chaotic at times, it has remained more consistent and predictable than has been the case in Canada where the uniformed military often has no idea what its policy (or acquisitions) will be next year, let alone five or ten years from now.

15   Gabriel Almond, Russell J. Dalton, G. Bingham Powell, Jr., Kaare Strom, *European Politics Today*, 4th edition (New York, Longman, 2010), p. 2.

of qualitative data, and the events that will be compared, were selected because they fit into these categories. In terms of time lines, the U.S. examples will begin with the Lyndon Johnson era, while the Canadian time line will begin with Paul Hellyer's restructuring of the Canadian military in the 1960s.

The key proposition advanced in this paper is that while conflict is ubiquitous, a failure to respect military culture will intensify conflict and undercut efforts to create "shared responsibility." The assumption is that it will lead to a less well-thought-out security policy. It should also be emphasized that the military is capable of committing blunders on its own: Canada's experience in Somalia and the American experience in trying to rescue the hostages in Teheran are obvious examples.

Canadian Civil-Military Relations

When he became minister of defense, Paul Hellyer decided to show the uniformed services, which had tended to avoid civilian leadership, that he was in charge. He refused to sign any documents during his first 30 days in office and suspended equipment purchases while a defense review was underway. He ignored the advice of senior military officers and proceeded to revamp the military's structure by unifying the services into a single force: Canadian Forces. This meant a new uniform, and new rank structure. In the process he ignored the military's past and tried to impose a civilian-style culture on the armed forces. He was successful in getting Bill 243 passed and the new face and structure of Canadian Forces began to appear.

Hellyer had little or no time for input from Canada's military. He was convinced that he knew what was best for it. It should come as no surprise that military opposition was rampant. The military would have been opposed to such changes regardless of who introduced them. But Hellyer made matters worse because of his disdain for senior military officers.

Examples of military resentment abound. For example, Rear Admiral William Landymore, mentioned previously, who commanded the Atlantic coast, was especially outspoken in his criticism. In a speech to sailors, he "left no doubt that he thought that unification would destroy the RCN and its effectiveness."[16] His firing by Hellyer was followed by a mass exodus from the military. For example, from January 1965 to August 1966, 28 general officers, including the Air Chief Marshal, who was Hellyer's first Chief of Defence Staff, left. In addition, all three of the "three stars" and a total of 79 senior officers resigned. A total of 26,300 other officers and senior NCOs also left. By August 1966, only two of the 13 most senior officers had been in their commands more than a month.[17] According to one

---

16   Ibid.
17   J.L. Granatstein, *Who Killed the Canadian Military?* (Toronto: HarperCollins, 2004), p. 79.

Canadian specialist, these men "resisted unification ... in ways that fit within the rules of Canadian civil-military relations."[18]

From the military's standpoint, Paul Hellyer is considered one of the country's worst defence ministers. He not only violated military procedure left and right, he belittled military tradition and culture. He spoke of "buttons and badges" in reference to military medals, which some had received at great personal cost. He cared not a whit for military culture.

Pierre Trudeau was not much better. In fact, he was probably one of the most anti-military of all of Canada's post-war prime ministers. He avoided military service during World War II, and openly stated that he "viewed soldiers as unintelligent thugs."[19] Trudeau quickly took charge of military affairs and in the process, he reduced Canadian Forces in Europe much to the chagrin of Canada's NATO allies, and cut the defence budget.

In the meantime, like many bilingual states, Canada had a problem creating a military that was open and accepted by both its francophone and anglophone citizens. Furthermore, unlike other changes he wished to make in Canada, Trudeau (and his successor) wisely left this one to the Chief of the Defence Staff, General Jean Allard. A combat veteran, Allard was one of the very few francophone officers in the upper ranks of the Canadian military and he was determined to improve the situation by pushing for bilingualism and equal treatment in the Canadian military. It was a struggle, but in June, 1969, the "Official Languages Act" was passed. While there was wide-spread opposition on the part of anglophone officers, it remained an intra-military issue (with support from the political world if needed). In the end, a quota system was imposed and it was at least partially successful. "Since 1970 the proportion of francophone officers had increased from 10.6 percent to 19.2 percent, and among the troops from 19.1 to 24.9."[20]

Trudeau's defence minister in the 1970s, Donald MacDonald, assumed leadership over the military and treated it much the same as Hellyer. He oversaw the drafting of a White Paper, which was released on August 24, 1971. He also appointed a Management Review Group to evaluate how the bureaucracy operated in the DND. While some Canadian officers were involved in the process, it was primarily run by MacDonald who made it clear that he was in charge. He was the leader. Besides, rather than work closely with the military, MacDonald brought a senior officer from External Affairs to work on the document.

The outcome was an organizational structure that clouded lines of responsibility. For example, it was unclear who sat at the top of the chain of command just below the minister. Who was accountable to whom? "In the meantime, accountability

---

18   Ibid.

19   J.L. Granatstein and Robert Bothwell, *Pirouette: Pierre Turdeau and Canadian Foreign Policy* (Toronto: University of Toronto Press, 1990), pp. 7-8, and Granatstein, *Who Killed the Canadian Military?*, p. 115.

20   Gerald Porter, *In Retreat: The Canadian Forces in the Trudeau Years* (Toronto: Deneau & Green, 1978), p. 25.

suffered because it became increasingly difficult to identify who was responsible for what decisions."[21] Not only were lines of authority blurred, so was an officer's job description. Canadian officers were visibly upset and complained about the civilianization of military policy as well as increased civilian interference in operational decisions. They believed it undermined the uniqueness of the CF and made it into just another form of public service.

The Fyffe Review was conducted under Joe Clark's short interregnum. Its goal was to produce a document to be entitled, *Canada in a Changing World*. Once again, military officers believed they were being ignored and that only civilians close to the minister of defence were being listened to. In the end, the report had little impact because the Trudeau government returned to office before it came out.

As far as the military was concerned, it had not only been starved for funds under Trudeau, but its organizational structure had been turned upside down. None of the goals that Hellyer had laid out had been fulfilled. Indeed, in the eyes of most of those in uniform, "the Hellyer plan was an unmitigated disaster for Canada's Armed Forces."[22] Fortunately for Canadian Forces, some changes were made—a new Task Force review group recommended that the three senior commanders became members of the Defence Council and Defence Management Group. For the first time, the military had a majority on both groups.

In spite of the improvements in military participation in decision-making noted above, the bottom line under Trudeau was that Canada's military was in a mess. Few politicians cared about it, and when a politician did pay attention to it—including ministers of defence—they seldom consulted the generals and took actions that increased their own political status. Note the following statement on the situation in 1983:

> Canada's military capabilities, once very significant, declined slowly throughout the Cold War era, but very serious retrenchments began around 1964 under the Liberal Government of Mike Pearson. Indeed the decline continued in the 1960s and then in the 1970s under Pierre Trudeau, when some major capabilities were eliminated entirely. By the early 1980s, the rustout of Canadian Forces was so pronounced and allied criticism so loud that the matter became a major issue in the 1983 election campaign.[23]

---

21   Douglas Bland, "The Government of Canada and the Armed Forces: A Troubled Relationship," in *The Soldier and the Canadian State: A Crisis in Civil-Military Relations?*, edited by David A. Charters and J. Brent Wilson (University of New Brunswick, Conflict Studies Workshop, 1995), pp. 28-9.

22   *Task Force on Review of Unification of the Canadian Armed Forces: Final Report*, March 15, 1980 (Ottawa: Ministry of National Defence, 1980).

23   Howie Marsh, "The Gathering Defense Policy Crisis," in Douglas Bland, *Canada Without Armed Forces?* (Montreal and Kingston: McGill-Queen's University Press, 2004), p. 85.

When the Mulroney Government came to power on September 17, 1984, the prime minister was determined to improve the situation and within three months had prepared a White Paper entitled *Challenge and Commitment: A Defense Policy for Canada*. It appeared to the military that help had finally arrived. At last, they had a government that was taking the military seriously. Unfortunately, the Mulroney government was faced with a fiscal crisis and the military budget was the first place to make cuts.

In 1986, a new defence minister Perin Beatty under Prime Minister Brian Mulroney concluded that Hellyer's decision to force all members of the Canadian Forces to wear the same uniform had been a failure. Much to the military's delight (especially the Navy), he ordered a return to the military's traditional uniforms. A year later, the service chiefs were reinstated and returned to NDHQ.

With the return of the traditional service chiefs the question arose of who was in charge? The three chiefs tended to see themselves as operational commanders. That made either the CDS or the service commander redundant. Once again no one knew who was in charge.

In the meantime, Canadian officers were shocked to see the 1989 defence budget. It had been drafted by the Prime Minister and the defence minister without their input. But the uniformed military were the ones who would have to implement it, and explain to their personnel why critical items were cut. This was a clear lack of respect. As one source put it, "Officers and soldiers were first surprised, then resentful, and finally sullenly resigned to the idea that no one valued their contribution, understood their needs, nor represented their points of view."[24]

Meanwhile, the service chiefs were gaining more and more authority and autonomy in areas such as promotion, or the army's staff college system. It had reached the point where it would be difficult, if not impossible to explain to an outsider where decisions were made.

It was also clear that Canada lacked a coherent security policy. In the meantime, Ottawa was getting CF involved in more and more peacekeeping operations around the world.

Just how bad the situation was inside the CF was became obvious when Ottawa sent peacekeeping forces to Somalia in 1992. After more than two months of consideration, the overloaded CF leadership decided to send an airborne regiment. Unfortunately, there were a whole series of problems with this unit: the troops were not trained for the operation, and there were questions about the suitability of some of the officers and troops. But this was all of the army that was ready. However, the CF was "reeling under the strain, its commitments vastly exceeding its capabilities."[25]

---

24 Douglas Bland, "The Government of Canada and the Armed Forces: A Troubled Relationship:" in *The Soldier and the Canadian State: A Crisis in Civil-Military Relations?*, edited by David A. Charters and J. Bent Wilson (Conflict Studies Workshop, University of New Brunswick, 1995), p. 32.

25 Garantstein, *Who Killed the Canadian Military?*, p. 152.

On March 16, 1993 several soldiers from the airborne regiment captured, tortured, and killed a Somali teenager who had been trying to steal something from them. Unfortunately, this action was covered up all the way to NDHQ in Ottawa. As the historian for the Canadian Airborne put it, "a conscious decision was made to control any political damage rather than see public justice done."[26]

The generals knew that they should never have sent the regiment to Africa, but the CDS General John de Chastelain wanted to please his political masters. Meanwhile, problems inside the CF were more wide-spread than in Somalia. A campaign carried out by non-commissioned officers "uncovered wide-spread fraud, deception and dereliction of duty in the officer corps."[27] The Prime Minister, Jean Chretien, would later set up an investigative committee to look at events in Somalia, but unfortunately it was headed by a judge who was unable to focus directly on the issues. According to Bercuson, it was a case of the vultures coming home to rest. As he put it:

> The crisis was caused initially by the deliberate bleeding of the defence establishment to near death by successive mostly Liberal, governments. It was made a great deal worse by unification and the imposition on Canadian Forces of a structure designed to ease political and bureaucratic burdens rather than the promise of military effectiveness. The sins of unification were compounded by the creation of National Defence Headquarters, designed to murder military initiative.[28]

Put simply, CF morale and moral compass was at its nadir. From the military's standpoint, Mulroney's heart was in the right place. He dealt directly with the senior military, but budget cuts made it impossible for him to rebuild CF.

Chretien was similar to Trudeau when it came to the military—he saw little use for it, and immediately cancelled the preceding government's order of 43 EH-101 military helicopters desperately needed by CF to replace 40-year-old Sea Kings. Then came the collapse of the Soviet Union—time for a peace dividend!

In 1994, the then defence minister, David Collenette, who did not believe the military was a special occupation, introduced a White Paper that outlined further cuts in defence spending. All of the hours of training, the hardships suffered by families and the troops with their long deployments whether on the ground, in the air, or at sea, meant nothing to him. This was an attack on the heart of military culture.

It was only a short time before Canadian forces were again sent abroad, this time to the former Yugoslavia. Ottawa demanded control of everything—almost all military actions had to be cleared by a civilian-dominated group in Ottawa. The word was "risk-averse." Avoid casualties! General Rick Hillier, who was in

---

26   Bercuson, *Significant Incident*, p. 224.
27   Douglas Bland, "The Government of Canada and the Armed Forces," p. 31.
28   Bercuson, *Significant Incident*, pp. 241-2. Emphasis in the original.

command of Canadian troops in the Balkans, said it best when he observed that the call-name given to the Canadian troops when they arrived was "Canbat 1" and "Canbat 2." After several months of the Canadians refusing this task or avoiding that job, their allies quickly changed their name to "Can't bat 1 and Can't bat 2."[29] This constituted a further attack on military culture—making it impossible to function as soldiers. Even the number of casualties was covered up.

Civil-military relations were at an all-time low. Bland described them as "floundering and uncertain."[30] By 2000, almost every piece of equipment in the CF inventory was worn out. Nevertheless, in the aftermath of 9/11, Canadian troops found themselves on their way to Afghanistan. On April 16, 2003, the Canadians agreed to take responsibility for the International Security Assistance Force in Afghanistan.

If any single individual can be singled out for helping bring about a major change in civil-military relations in Canada, it was General Hillier who became CDS. He immediately began making major changes—with civilian approval. As Stein and Lang observed, "in a few weeks' time, a soldier from Newfoundland had outclassed and outrun the best minds in Canada's august Department of Foreign Affairs."[31] He was determined to show the world that the Can't Bats in Bosnia had become the Can Bats in Afghanistan. He also made it his job to convince civilian leaders that the CF was over-burdened with too many commitments. Prime Minister Paul Martin paid Hillier one of the highest compliments a solider can receive:

> Hillier's appointment would fundamentally change the philosophy, the strategy, the organization and the culture of Canadian Forces. He would become the most important and influential CDS in living memory … Hillier would make defence policy his first priority. Defense policy was historically and quite appropriately, the domain of civilian officials, and it was unprecedented for a chief of defence staff to be given this responsibility.[32]

He also quickly came up with a plan for increasing Canada's role in Afghanistan, an action that impressed the American Pentagon.[33] Hillier also pushed another key component of military culture, the wide-spread used of traditional symbols. As he put it, "This rebirth of tradition signaled the rebirth of pride."[34] In addition, he was

---

29   Rick Hillier, *A Soldier First: Bullets, Bureaucrats and the Politics of War* (Ontario: HarperCollins, 2009), p. 158.

30   As cited in English, *Understanding Military Culture*, p. 57.

31   Janice Gross Stein and Eugene Lang, *The Unexpected War: Canada in Kandahar* (Toronto: Viking, 2007), p. 157.

32   Paul Martin, *Hell or High Water: My Life in and Out of Politics* (Toronto: Emblem, 2009), p. 151.

33   Stein and Lang, *The Unexpected War*, p. 181.

34   Hillier, *A Soldier First*, p. 335.

the force behind the 2005 Defence Paper which became an "integral part of the government's International Policy Statement, rather than a separate government white paper."[35] Hillier believed it was time for Canada to realize that the primary task for CF was to support Canadian foreign policy, not just by being overworked peacekeepers, but by having modern weapons that would stand behind a strong foreign policy.

Hillier believed in the concept of a shared responsibility. As he put it in an interview in 2007, "The Government of Canada sets the agenda and it is our job to provide military advice and counsel on that agenda and how it can affect Canada's military capability. I either support it or, if I am at significant odds with that agenda, I resign."[36] As Hillier noted, however, there would not have been shared responsibility without the support that PM Paul Martin provided. For example, he immediately went to visit defence headquarters after taking office, which he said was a "signal of the importance I placed on our Armed Forces."[37] The military respected him even if he was not able to deliver on his promises to rebuild CF.

Stephen Harper, Martin's successor was also convinced of the need for a strong military to back up Canada's foreign policy. Unfortunately, at the same time Harper was making clear his commitment to modernizing the CF, the war in Afghanistan was demanding more and more resources. Hillier himself admitted that he underestimated the costs involved: "I underestimated the demands of the Afghan deployment."[38]

Hillier's most important impact was the reinvigorated CF. As one source put it, "After more than a decade of our military being mislabeled as mere peacekeepers, Hillier has made the rank and file proud to proclaim themselves soldiers again."[39]

On May 12, 2008, Hillier's successor, General Walter Natynczyk announced the Canada First Strategy (originally formulated under Hillier's leadership).[40] Regardless of how much of it is actually implemented, it marks a major shift from the days of civilian domination of the military with little attention paid to its culture to one of civilian supremacy with greater attention now given to military views and advice. In other words, towards at state of shared responsibility. There are problems on the horizon, however. For example, on October 11, 2010, Hillier warned against civilian involvement in the military command process which

---

35   Daniel Gosselin and Craig Stone, "From Minister Hellyer to General Hillier: Understanding the Functional Differences Between the Unification of Canadian Forces and its Present Transformation," *Canadian Military Journal*, 6:4, http://www.journal.dnd.ca/vo6/no4/trans-eng.asp/

36   "Q&A: General Rick Hillier on Canada's Mission in Afghanistan," *The Globe and Mail*, March 12, 2007.

37   Martin, *Hell or High Water*, p. 393.

38   Stein and Lang, *The Unexpected War*, p. 243.

39   "O'Connor Lacked Hillier's Charm," *Esprit de Corps*, 2009, http://www.espritdecorps.ca/index.php?option-com_content&view-a/

40   "Q&A: General Rick Hillier on Canada's Mission in Afghanistan."

he warned could undermine the military command and control process.[41] In the meantime, by April 2011, Canada was locked in an election campaign. Given past history, the winner will have a major impact in the nature of Canadian civil-military relations.

The U.S.[42]

Of all of America's post-World War II presidents, no one played a more negative role in American civil-military relations than President Lyndon B. Johnson. Like Trudeau, he made no secret of his contempt for the military, believing that most of them were "arrogant" and that they were contemptuous of new ideas. As one writer put it, "Johnson could be merciless when he talked about the generals."[43]

Robert McNamara, his Secretary of Defense, was no better. He brought with him a group of civilian experts on security affairs (the Whiz Kids) and in most cases used them as a substitute for the uniformed military when it came to drafting plans on security related issues. For example, rather than asking the Joint Chiefs for their suggestion for a strategy for the war in Vietnam, he forced his ideas on them. He came from the Ford Motor Corporation and did his best to civilianize decision-making in the Pentagon. His plan called "not for imposing his will on the enemy, but to communicate with him."[44] The military believed the U.S. should get in with both feet or get out, but McNamara had no time for them or their ideas. As General Andrew Goodpaster put it to McNamara with reference to the war in Vietnam, "Sir, you are trying to program the enemy and that is one thing that we must never try to do. We can't do his thinking for him."[45] Goodpaster's words fell on deaf ears. Indeed, McNamara made it clear to the Chiefs that their advice was unwelcome.

For the military, the situation was so bad that the Chiefs came very close to resigning. They eventually decided not to resign at the last minute. However, it was clear that civil-military relations had reached a new low point. In the future, their meetings with McNamara "were ruled by an icy cordiality that conformed strictly with the requirements of courtesy, but no more."[46]

On the surface, Richard Nixon would seem to be just the kind of president the military was looking for in the aftermath of Johnson's time in office. However, Nixon attempted to control everything that went on—and that included the

---

41   "Hillier Warns Against Civil Servants Directing Military Operations," *The Globe and Main*, October 11, 2010.

42   For reasons of space, this discussion begins with a discussion of Lyndon Johnson's presidency and the role played by both Johnson and Robert McNamara.

43   Hugh Sidney, *A Very Personal Presidency* (New York: Atheneum, 1968), p. 203.

44   Lawrence Freedman, *Kennedy's Wars* (Oxford: Oxford University, 2000), p. 40.

45   H.R. McMaster, *Dereliction of Duty* (New York: Harper, 1998), p. 163.

46   Mark Perry, *Four Stars* (Boston, MA: Houghton Mifflin, 1989), p.165.

Pentagon; he was also extremely secretive, and manipulative. In addition, he refused to follow the chain of command. This put the Chiefs in a difficult position. They were trained to follow the chain of command, but Nixon (or his National Security Advisor Henry Kissinger) would approach them directly telling them not to inform either the Chairman of the Chiefs or the Secretary of Defense of the president's plans or ideas. One officer, Admiral Elmo Zumwalt, when asked by Kissinger to set up a back-channel communications system, refused and informed the Chairman, Admiral Thomas Moorer. On the other hand, Secretary of Defense Melvin Laird respected them and dealt honestly with them, even though both Kissinger and Nixon told senior officers that they did not trust Laird.

Nixon talked tough on Vietnam. He came into office stating that he had a "secret plan," when no such plan existed. He was determined, however, to show Hanoi and Washington that he was tough. While they appreciated his more aggressive approach, the Chiefs were concerned about the ethical nature of his back-channel actions and his secrecy—going to the point of forcing the Air Force to make up two sets of bombing plans—one public showing no bombing of Cambodia, while the other (classified) showed that the U.S. was bombing Cambodia.

Meanwhile, the Chiefs retaliated by making use of a Navy Yeoman, who was assigned to the White House JCS Liaison Office. He copied secret documents and sent them to the JCS. It demonstrated that the President was so secretive that they had no alternative but to employ a spy in the White House in order to find out what was happening.

The military's disloyalty *vis-à-vis* Laird became obvious, when in a meeting with the President, Admiral Moorer openly disagreed with his boss, telling Nixon with regard to Operation Linebacker that the Chiefs "were ready to walk out the door unless Nixon did it."[47]

One of the clearest signs of how little respect Nixon had for military culture was his dealing with Zumwalt who had been promoted to the Chief of Naval Operations. There had been a number of racial incidents bad enough that sailors had to be evacuated from a ship to onshore hospitals. The primary instigators were black, and Nixon wanted them out of the Navy—immediately. The problem was that Zumwalt did not have such authority. The *Uniform Code of Military Justice* required that the sailors involved go through a lengthy legal process. Kissinger shrieked at Zumwalt demanding that he give them dishonorable discharges, but Zumwalt refused. He then told Zumwalt, "You should be aware that the President thinks that the JCS are 'a bunch of shits' and that you are the 'biggest shit of all.'"[48] It was anything but shared responsibility.

Jimmy Carter was an unknown.[49] On the one hand, he was a graduate of the U.S. Navy Academy and clearly understood military culture. On the other,

---

47   Perry, *Four Stars*, p. 209. In fact, Moorer made this up. The Chiefs had said no such thing.

48   Elmo R. Zumwalt, *On Watch: A Memoir* (New York: Quadrangle, 1976), p. 506.

49   Gerald Ford's short time in office is not covered for reasons of space.

throughout his campaign he focused on domestic problems, while the military was close to collapse. In addition, Carter was difficult to deal with. He did not believe in structure and often no one knew what the President meant or who was in charge of implementing his decision. In addition, he was a micro-manager.

The Chiefs' first meeting with him was a disaster. They went through their normal briefing, with graphs, slides and explanations—all aimed at convincing the President just how bad the situation was inside the armed forces. After their presentation, Carter looked at them and said, "How long would it take to reduce the numbers of nuclear weapons in our arsenal? He followed up by asking, "What would it take to get down to a few hundred?"[50] The Chiefs were stunned. As far as the budget was concerned, Carter was prepared to make massive cuts and did not care what they thought.

Carter was serious. He attempted to pull American troops out of Korea, to cancel the B-1, and also to cancel construction of the Navy's new 90,000-ton aircraft carrier. These actions together with his willingness to cut the budget led to a situation where "seldom in the memory of Pentagon observers did the generals split with the President on the military budget, as they did in an understated way today."[51]

One of the few instances in which Carter showed respect for military culture was the Iranian Rescue attempt. He left the planning and the operation in the hands of the military. As Carter told General David Jones, "David, this is a military operation and you're going to run it."[52] Most important, when the mission failed, Carter lined up the survivors and took full responsibility for the mission's failure just as a senior military officer would.

When Ronald Reagan came into office he promised to restore morale and give the military the respect it had enjoyed in the past. Given what they had been through under Nixon and Carter, Reagan seemed like manna from heaven. The only downside for the military was Reagan's chaotic decision-making style. He gave the orders (which were often ambiguous) and left it to the military to implement them. This ran in the face of a key aspect of military culture—the desire for predictability. On the other hand, it was clear that he was prepared to pour money into the armed forces. Secretary of Defense Caspar Weinberger understood little about the military as he demonstrated during the Chief's first meeting with him. The lesson for the Chiefs was simple: humor the Secretary when he made such suggestions while at the same time praising him for the money he brought in. As far as the increase in money was concerned, Colin Powell probably put it best when he said that "This was Christmas in February. This was tennis

---

50   Perry, *Four Stars*, pp. 265-7.

51   "Joint Chiefs Dissent on Carter-Brown Military Budget," *New York Times*, May 30, 1980.

52   Charles Beckworth, *Delta Force* (New York: Avon, 1983), p. 249.

without a net. The Chiefs began submitting wish lists. The requests initially totaled approximately a nine percent increase in defense spending."[53]

The military was clearly pleased with Reagan's decision to open the floodgates when it came to money to fix the military's myriad problems. The problem was that these decisions were not being discussed with the military. The generals knew this kind of largesse could not last forever and would have been much happier with less money, and a more carefully crafted budget.

Reagan did not always heed the military's advice. When it came to sending troops to Lebanon, the Chiefs were uniformly opposed to it. In fact, marines had no sooner landed when the situation began to fall apart. Less than a month after they landed, the marine barracks was blown up and a total of 241 marines, sailors and soldiers were killed.[54]

The situation was different with Operation Fury (Grenada). Reagan was directly engaged in preparations for it. The Chiefs presented their plan to seize the island and Reagan went along with it. However, when it came time to send troops ashore, it was the Pentagon that called the shots. There were problems with the operation because of fights between the services (primarily Army and Marine Corps).

An effort to improve military cooperation came with the Goldwater-Nichols Defense Reorganization in 1986. It required senior officers to serve at some point with another service, and made the Chairman of the Joint Chiefs of Staff the primary military advisor to the secretary of defense and the president. Despite his sometimes chaotic leadership style, the Chiefs respected Reagan and did their best to work around times when he violated military culture well aware that he had their best interests at heart. While the relationship might not have reached the point where it could be called a shared relationship, it was certainly far healthier from the military standpoint than was the case with his recent predecessors.

George Bush played a greater role in national security decision-making than did his predecessor. He did not micro-manage the process, but he expected to be consulted on important issues. When it came to major decisions, Bush was more pragmatic than his predecessor and he also made clear exactly what he expected from his subordinates.

The first major military incident facing Bush was the invasion of Panama. Faced with Panamanian President Manuel Noriega's brutish behavior toward American military personnel, Bush and the Chiefs decided that something had to be done. The decision was to invade Panama. During the planning for the invasion and the actual deployment of troops, Bush was fully informed of what steps were being taken by Colin Powell, but he left the details to the military. According to Powell, after having the plan laid out, Bush said, "Okay, let's do it," "The Hell with it."[55]

---

53   Colin Powell, *My American Journey* (New York: Random House, 1995), p. 258.
54   Perry, *Four Stars*, p. 312.
55   Powell, *My American Journey*, p. 425.

Bush's behavior during the preparation and carrying out of the Persian Gulf War was similar. He was fully briefed on all plans and he made the final decision when Allied troops should stop their movement into Iraq. Powell played a very important role as an interface between the generals in the field and the President and Secretary of Defense, but the military very much appreciated Bush's support, encouragement and refusal to try to micro-manage matters. Indeed, from a military standpoint Bush was the ideal president. He was in charge, gave clear instructions, and supported the generals and the troops. This was a shared responsibility at its best.

William Clinton did not start off well with the military. He was the first U.S. president since World War II who had not served in the military, and as a result, understood almost nothing about military culture. His leadership style was upsetting. One scholar called it "Ad hocracy in action."[56] He often equivocated and then made a decision on the spur of the moment. They also resented his decision to use the military as a laboratory for social engineering and felt that he did not respect them. After all, at one point, he had stated that he "loathed the military."[57]

Then there was case of homosexuals serving in the military. Clinton promised during his campaign that he would lift the ban on their service. Once in office, he prepared to lift it only to find strong opposition not only on the part of Colin Powell and the Chiefs, but by several senior senators as well. In the end, he was forced to make the well-known "don't ask, don't tell" compromise. In addition, he did not follow foreign-policy issues closely and every time an incident occurred that caught him off guard, he would blame his advisors, including those in the military. He was never responsible for U.S. actions (for example, Somalia, Haiti, Bosnia). This led one distinguished retired military officer to exclaim, "I don't know which is worse: the attempt to duck responsibility or the image of a Commander in Chief so out of touch that he allowed his foreign policy and military aims to take entirely different paths—both fatally mined."[58] While senior military officers always found him polite and willing to listen, the relationship was difficult. There was no leadership as far as they were concerned.

For the military, George W. Bush seemed a major improvement over Clinton, but they soon ran into a problem and it was his Secretary of Defense, Donald Rumsfeld. Rumsfeld expected to be in the center of all military-related decisions. Indeed, he had little time for the uniformed military or their advice and counsel. He carefully installed an officer as Chairman of the Joint Chiefs, General Richard Myers, who he was convinced would do his will—and he did.

---

56   John Burke, *The Institutional Presidency Organizing and Managing the White House from FDR to Clinton*, 2nd edition (Baltimore, MD: Johns Hopkins University Press, 2000), p. 180.

57   George Stephanopoulos, *All Too Human* (Boston, MA: Little, Brown, 1999). p. 75.

58   Cited in Elizabeth Drew, *On the Edge: The Clinton Presidency* (New York: Simon & Schuster, 1994), p. 358.

Rumsfeld was convinced that the U.S. military was not doing enough to adapt to modern technology. His campaign focused on the Army, in spite of the fact that the Army Chief of Staff, General Eric Shensiki had already begun his own effort to transform it. Over the next several years he would insult Shensiki and ignore military advice. He interfered not only in the military's plans to invade Iraq, but in operations as well. For example, he continually badgered General Tommy Franks over the number of troops until he got it from 385,000 (the number Franks' predecessor, General Tony Zinni had recommended), to around 150,000. Then phase four in the military's operational plans—post-combat operations—was completely ignored, which together with the paucity of American troops was a major contributing factor to the insurgency that developed in Iraq.

Rumsfeld continued to alienate the uniformed military. His plans for the post-invasion period did not work, and it was not long before he and NSC advisor Condoleezza Rice locked horns over American policy *vis-à-vis* Iraq. Indeed, by 2006, he was being openly criticized by retired U.S. military officers. As far as the uniformed military was concerned, he ranked with Robert McNamara as one of the worst Secretaries of Defense. There was no thought of joint responsibility.

His replacement, Robert Gates, was a very different individual. He was a strict disciplinarian. He believed strongly in one aspect of military culture— accountability. He fired officers who he believed were not doing their duty such as senior Army officers because of the miserable conditions in the Walter Reed Army Hospital in Washington, D.C. or Air Force generals whom he held responsible when nuclear weapons were not handled correctly. But this type of behavior did not bother the military; he was fair and just.

Senior military officers also found him easy to work with on policy toward hot spots such as Iraq and Afghanistan. This was also the case under Barack Obama. He made sure the President understood their views even when they disagreed with his. This is especially evident in Bob Woodward's book on planning in the White House for an increase of U.S. troops in Afghanistan.[59] The military was given a full opportunity to present its information, and there were heated discussions on what the U.S. should do. Admiral Mike Mullen and General David Petraeus played prominent roles but it was clear throughout the book that President Obama had the final word in such matters—another case in which there was a special relationship.

Conclusion

The purpose of the synopsis was to argue that in spite of the inevitable conflict that is part of civil-military relations, much of it can be ameliorated depending on how civilians treat the military. If they show respect for military culture they have a much better chance of attaining the "shared responsibility" mentioned above.

---

59   Bob Woodward, *Obama's Wars* (New York: Simon & Schuster, 2010).

Looking at the Canadian experience, it appears that it is only in recent years, beginning with Brian Mulroney, but really developing under Paul Martin, and now continuing under Stephen Harper, that there has been increasing respect for military culture. As noted above, General Rick Hillier's magnetic personality and drive also played a major role and his successor General Walter Natynczyk appears to be playing a less public but positive role in interacting with his political masters.

On the other hand, civilian officials like Hellyer, Trudeau, Cretien and some of their ministers of defence exacerbated conflict because of their refusal to pay any attention to military culture regardless of the reason behind it. The period from Hellyer to Mulroney appears to have been a very negative one for Canadian Forces; a situation made worse by the disdain shown generally by civilians for the military. Likewise, the tendency on the part of prime ministers to continue sending exhausted, ill-equipped and undermanned CF on peacekeeping missions all over the world showed lack of concern for the troops, something that is a critical part of military culture.

The U.S. military always had an advantage over Canadian Forces for the reasons noted above.[60] However, it too has seen an ebb and flow in civil-military relations. Conflict under Lyndon Johnson and Robert McNamara was rampant. The same was true of Donald Rumsfeld. Jimmy Carter created problems, not only because he insulted military culture but because he starved the armed forces. Likewise, the military appreciated Reagan for two reasons. First, he respected military culture, and in spite of Weinberger's efforts to get in the way, there was a "shared responsibility" when it came to military operations such as Grenada and Lebanon, even when the generals disagreed with the president. Second, Reagan poured money on them; however, in this case, they wished that the largesse they were receiving was better planned.

From the generals' and admirals' point of view George Bush was the President who respected military culture the most. When he disagreed with the brass, he did so in a respectful fashion. Bill Clinton was a problem because he not only refused to accept responsibility, a key aspect of military culture, but his decision-making style often left the generals and admirals confused as to his intent.

George W. Bush's problem was his Defense Secretary. When he was replaced by Robert Gates, the relationship with the military improved significantly even though he was tough on those who failed to live up to his standards. Finally, Barack Obama appears to have an excellent relationship with the uniformed military.

So where does this leave us conceptually? This approach is impressionistic, not the kind that lends itself to quantification. Its advantage is that it focuses on the actors; both military and civilian and does not assume that they live in separate worlds. There is no guarantee that a shared responsibility will necessarily lead to good military advice. Humans are imperfect, including generals and admirals. Furthermore, the goal is not to create a civilian elite who will always give in to

---

60   See footnote 15 above.

military demands. Civilians are in charge both in Canada and the U.S. and both militaries know and accept it.

Several years ago, the late Richard E. Neustadt commented that "politics is the art of persuasion."[61] In a sense that is what this chapter is arguing. When militaries are controlled by civilians as both of the two discussed here are, civilians will get a lot more out of them if they understand that military personnel come from a different culture and respect it (not adopt it). In respecting the way military officers think and act, they are not lessening civilian control; rather they are following Neustadt's advice and making civil-military relations the art of persuasion.

---

61   Richard Neustadt, *Presidential Power and the Modern President: The Politics of Leadership from Roosevelt to Reagan* (New York: Free Press, 1990), p. 11.

Chapter 3

Civil-Military Relations in Contemporary Russia

Stephen J. Blank

The state of civil-military relations is a highly reliable barometer of the nature of the government and its overall relation to its people. Recent scholarship on Russian civil-military relations explicitly confirms this point. Thomas Gomart writes that:

> Through the civil-military relationship the nature of a state's politico-strategic project can be assessed, that is, what is its understanding of the world; what resources does it have available, what is its willingness to modify its international environment. Studying the civil-military relationship also makes clear current modes of power, the sharing of responsibility in security matters, and in certain cases the will to act.[1]

Similarly Zoltan Barany writes that:

> The reform of the armed forces is closely connected, through the broader issues of civil-military relations, to the general state of Russia's democratization. The politics of defense reform is at the core of Russia's democratization given the crucial role the military establishment has played throughout Russian history, including the more than seven decades of Communist rule during which the Soviet Union had built a great military empire.[2]

An analysis of these issues under conditions of defense reform and the current international situation is of immense analytical and policy relevance for both domestic and external security in Russia. But its conclusions of such an inquiry are utterly depressing. The current turmoil in the Middle East underscores the fact that a government's control over its armed forces and the latter's relationship to society are often the ultimate argument in both domestic politics and questions of war and peace. Russia's profile on these issues unfortunately resembles in many

---

1   Thomas Gomart, *Russian Civil-Military Relations: Putin's Legacy* (Washington, D.C.: Carnegie Endowment for International Peace, 2008), p. 87.

2   Zoltan Barany, *Democratic Breakdown and the Decline of the Russian Military* (Princeton, NJ: Princeton University Press, 2007), p. 177.

ways that of these Arab states and other third-world governments as shown by the following phenomena.[3]

- There is a systematic absence of democratic controls or accountability throughout the system. Despite legislation, neither the government, nor its members, nor officers or other members of the armed forces are effectively accountable to the rule of law or to parliamentary scrutiny. Law is a measure by which the strong punish the weak or as Vladimir Putin would say it is an instrument of dictatorship, clearly an extra-legal form of rule.
- No regularized system for formulating or implementing defense policy exists. We see here the rule of men not institutions, let alone law. The Security Council's purview and powers depend completely upon the whims of the topmost leadership and the General Staff and Ministry of Defense are perennially in conflict with each other even though the General Staff is legally subordinate to the Minister.[4] This process has apparently continued unabated since the late Soviet period.[5] As a result policy emerges out of an unstructured process dominated by personalities and not based on any concept of regularized government or what early modern German philosophers called a *Rechtstaat*, that is, a state based on law.
- As Zoltan Barany has pointed out, the government since 1990 deliberately encouraged the politicization of the armed forces.[6] Consequently as he shows the officer corps and organizations linked to it have deliberately fostered the political activism of men in uniform and the government has used the armed forces (both the army and police) for domestic, constabulary, and repressive measures against dissidents for 20 years.[7] When Defense Minister Anatoly Serdyukov, supported strongly by President Medvedev and Prime Minister Putin, launched his defense reform in 2008, enormous military and public opposition developed and still continues. Thus serving

---

3   Stephen J. Blank, *Civil-Military Relations in Medvedev's Russia* (Carlisle Barracks, PA: Strategic Studies Institute, U.S. Army War College, 2011).

4   Stephen J. Blank (ed.), *Russian Military Politics and Russia's 2010 Defense Doctrine* (Carlisle Barracks, PA, 2011) shows how this rivalry played out in the process leading to the formulation of the 2010 doctrine.

5   Dale R. Herspring, *The Soviet High Command, 1967-1989: Personalities & Politics* (Princeton, NJ: Princeton University Press, 1990).

6   Barany, passim.

7   Ibid.; Stephen J. Blank, "The Soviet Army in Civil Disturbances, 1988-1991," *The Military History of the Soviet Union,* edited by Robin Higham and Frederick W. Kagan (New York: Palgrave, 2002), pp. 275-98; Stephen J. Blank, "Civil-Military Relations and Russian Security," in *Civil-Military Relations in Medvedev's Russia,* edited by Stephen J. Blank (Carlisle Barracks, PA: Strategic Studies Institute, U.S. Army War College 2011), pp. 1-76.

troops demonstrate openly against the reforms.[8] Furthermore serving officers, for example Captain First Rank Sergei Gorbachev, tell an interviewer that:

> The situation which led to the appearance of many problems indicates that there is no clear concept at the top of the country's military-political leadership as to what the Armed Forces must be like. Although loud, fine statements are made and money allegedly is being allocated to resolve social problems, there is nothing visible that can be felt for now. Alas this is a fact.[9]

- Legislation permits Gazprom and Transneft to have private or semi-private military or quasi-military forces calling into question the monopoly of legitimate violence by the state.
- Meanwhile due to its fear of Russia's citizens' potential for demanding reforms and democracy as in the color revolutions of 2003-2005, the Russian government has resorted systematically to the creation of multiple militaries, including the regular armed forces, MVD, FSB, Border Troops, and so on. Indeed the MVD evidently got more funding till 2007 than did the regular Army, suggesting the real center of gravity of the regime's threat perceptions.[10] The extent of creation of new paramilitary forces or of adding domestic missions to existing structures eloquently testifies to this fact. An April 2009 report outlined quite clearly the threat perceived by the authorities. Specifically it stated that:

> The Russian intelligence community is seriously worried about latent social processes capable of leading to the beginning of civil wars and conflicts on RF territory that can end up in a disruption of territorial integrity and the appearance of a large number of new sovereign powers. Data of an information "leak," the statistics and massive number of antigovernment actions, and official statements and appeals of the opposition attest to this.[11]

---

8   For an example of such protests by the Union of Airborne Troops of Russia see Moscow, *Ekho Moskvy Radio*, in Russian, February 23, 2011, *Open Source Center, Foreign Broadcast Information Service, Central Eurasia* (Henceforth *FBIS SOV*), February 23, 2011.

9   "Interview With Captain 1st Rank, Sergey Gorbachev, Moscow, *Novyy Region*, in Russian, February 15, 2011, *FBIS SOV*, February 18, 2011 Whether or not Gorbachev's arguments are correct the key point is that he made them publicly while serving.

10   Julian Cooper, "The Funding of the Power Agencies of the Russian State," *Power Institutions in Post-Soviet Societies*, 6-7 (2007), www.pipss.org.

11   "Russia on the Brink of Civil War," Moscow, *Vlasti*, in Russian, April 19, 2009, *FBIS SOV*, April 19, 2009.

These agencies expected massive protests in the Moscow area, industrial areas of the South Urals and Western Siberia and in the Far East while ethnic tension among the Muslims of the North Caucasus and Volga-Ural areas is also not excluded. The author also invoked the specter that enraged former Army officers and soldiers who are now being demobilized because of the reforms might also take to the streets with their weapons. But despite this potential unrest, the government characteristically resorted to strong-arm methods to meet this threat, thereby repeating past regimes (not least Yeltsin's) in strengthening the Ministry of Interior's forces (VVMVD) and other paramilitary forces as well.[12]

This report and other articles outlined the ways in which the internal armed forces are being strengthened. Special intelligence and commando subunits to conduct preventive elimination of opposition leaders are being established in the VVMVD. These forces are also receiving new models of weapons and equipment, armored, artillery, naval, and air defense systems; 5.5 billion rubles were allocated in 2008 for these forces' modernization. Apart from the already permitted "corporate forces" of Gazprom and Transneft the MVD is also now discussing an *Olimpstroi* (Olympics Construction) Army and even the Fisheries inspectorate is going to create a special armed subunit called Piranha.[13]

As of 2003, there were 98 special-purpose police detachments (OMONs) in Russia. By comparison in 1988 during the crisis of the regime and its elites under Gorbachev 19 OMONs were created in 14 Russian regions and three union republics. By 2007, there were already 121 OMON units comprising 20,000 men operating in Russia. Moreover by 2007 there were another 87 police special designation detachments (OMSNs) with permanent staffing of over 5,200 people operating with the internal affairs organs, making a grand total of 208 special purpose or designated units with 25,000 well-trained and drilled soldiers. The OMSVs have grown from an anti-crime and anti-terrorist force to a force charged with stopping "extremist" criminal activity. All these units train together and have been centralized within the MVD to fight "organized crime, terrorism, and extremism." From 2005 to 2006 the financing of these units was almost doubled. By 2009, they were also working with aircraft assets, specifically the MVD's own Aviation Center with nine special-purpose air detachments throughout Russia. Seven more such units are to be created. Furthermore the MVD has developed a concept for rapidly airlifting these forces to troubled areas from other regions when necessary. These forces are also receiving large-scale deliveries of new armored vehicles with computers in some cases and C3 (command, control, communications) capabilities. Since these are forces apart from the regular VVMVD, "On a parallel basis with the OMON empire, a multi-level internal security troop machine is being developed-with its own special forces, aircraft, armored equipment, situational-crisis centers, and so forth."[14] These trends

---

12   Ibid.

13   Ibid.

14   Iriana Borogan, "In Shoulder-Boards: The Kremlin's Anti-Crisis Project: When OMON Rushes to Help," Moscow, *Yezhenedevnyi Zhurnal*, in Russian, December 15,

clarify why already in 2008 Russia announced that it would increase funding for the Ministry of Interior by 50 percent in 2010 and where the government's estimation of the true threat to Russian security lies.[15]

- All these multiple militaries and the state are wholly corrupted and criminalized In 2010, corruption cases rose by over 40 percent coming to almost 60,000 in the state as a whole.[16] Despite President Medvedev's three-year long campaign against corruption, preliminary signs show that despite now publicized efforts to uproot corruption and crime in the armed forces as a whole, again to include the VVMVD, in fact, despite the reform, the incidence of such events is rising. If this means reporting has improved that is a welcome sign. But the current anti-corruption campaign has yet to land any truly big fish and in many ways reflects more the settling of scores atop the government machine than a commitment to observing the law.[17]

Ultimately the pervasive corruption and criminality that we see in those forces reflect larger trends in the society and states a whole. These incidences of corruption, lawlessness, criminality, and aggressiveness are profoundly significant because they can have wide-ranging, unpredictable, and even dangerous consequences for Russia and its overall policy, both domestic and foreign, that can add considerably to the already considerable number of security challenges that Russia both presents to us and that it also perceives. For example it should be noted that when the Spanish police broke open a Russian Mafia mob in Spain in 2008, it turned out that the "Capos" of the Russian Mafia there were closely and personally tied to some of the highest ranking officials in the Russian government, for example, winning lucrative public works contracts. Yet they also clearly had contacts with terrorists in the North Caucasus.[18] This should not be surprising as former Minister of the Interior Anatoly Kulikov was already warning by 2005 of the criminalization of the state and the fusion of criminal and state organizations.[19] Likewise it is clear

---

2009, *FBIS SOV*, December 15, 2009.

15   Moscow, *Agentstvo Voyennykh Novostey Internet Version*, in Russian, July 4, 2008, *FBIS SOV*, July 4, 2008.

16   "Corruption Cases Rose by over 40 percent in 2010, Chief Russian Investigator," *Interfax*, February 18, 2011.

17   See for example, Stephen J. Blank, "The Putin Succession and Its Implications for Russian Politics," *Post-Soviet Affairs*, XXIV, 3 (September 2008): 231-62.

18   Madrid ABC.es, in Spanish, July 3, 2008, *FBIS SOV*, July 3, 2008.

19   Anatoly S. Kulikov, "Organized Crime in Russia: Domestic Developments and International Implications," *Russia and Europe: Putin's Foreign Policy*, edited by Yuri Fedorov and Bertil Nygre (Stockholm: Swedish National Defense College, 2005), pp. 115-28.

that corruption, indolence, and neglect have facilitated the commission of terror attacks in Russia, for example, the Domodedovo airport attacks in early 2011.[20]

Corruption within the security sphere this issue became prominent due to the audit conducted by Defense Minister Sergei Ivanov in 2006-2007 that revealed that corruption was even worse than expected. For example, on April 3, 2008 the Audit Chamber announced that more than 164.1 million rubles had been stolen from the ministry through fraud and outright theft. Another report stated that the Ministry of Defense (MoD) "accounts for 70 percent of the budgetary resources used for purposes other than those officially designated."[21] But while President Putin recognized the need for a new broom to sweep clean the Ministry and appointed Anatoly Serdyukov to do so, it is clear that despite Serdyukov's best efforts corruption continued and is still going on. Thus, "The Russian budget lost more than 6.5 billion rubles last year because of corruption in the country's Armed Forces, Russia's Chief Military Prosecutor Sergei Fridinsky said Thursday …"[22]

Similarly a 2009 audit revealed significant violations of financial and economic activity in the Air Force, amounting to a loss of over 660 million rubles. These violations occurred in the use of Air Force resources and funds by officials in Air Force commands, military units and organizations.[23] Thus, this corruption pervaded the Air Force. And this pervasiveness embraces as well the entire armed forces not just the Air Force. In 2008, Russia's leading defense correspondent, Alexander Golts, told a U.S. audience that 30-50 percent of the annual defense spending in Russia is simply stolen.[24] Subsequently prosecutors uncovered mass fraud in Rosoboronzakaz (Russian State Defense Purchasing Agency) in the amount of 6.5 billion rubles as well as the unlawful spending of 1.3 billion rubles and the inappropriate use of funds of 98 million rubles.[25] From January-August 2009 alone an investigation uncovered 1,343 violations of the law on the placement of defense orders in Rosoboronzakaz alone.[26] Indeed, an earlier investigation in June by the

---

20   Amy Knight, "Why the Kremlin Can't Fight Terrorism," *New York Review of Books Blog*, February 16, 2011, http://www.nybooks.com/blogs/nyrblog/2011/feb/16/why-kremlin-cant-fight-terrorism/

21   "Defense Ministry Will Shed Excess Equipment," RFE/RL Newsline, April 3, 2008, http://www.rferl.org/content/Article/1144084.html. See also, "Russian Official Says 30 Percent of Military Budget Lost Through Corruption," *Agentstvo Voyennykh Novostey*, July 2, 2008, also in *World New Connection* (hereafter as WNC) (articles available by subscription, see http://wnc.fedworld.gov/index.html).

22   "Corruption in Army Cost Russia Over 6.5 Billion Rubles in 2010 – Prosecutor," *Interfax*, February 24, 2010.

23   Moscow, *RIA OREANDA*, in Russian, July 1, 2009, *FBIS SOV* August 17, 2009.

24   Remarks by Stephen J. Blank, Eugene Rumer, Mikhail Tsypkin, and Alexander Golts at the Heritage Foundation Program, "The Russian Military: Modernization and the Future," April 8, 2008, http://www.heritage.org/press/events/ev040808a.cfm/

25   Moscow, *RIA Novosti*, in Russian, August 31, 2009, *FBIS SOV*, August 31, 2009.

26   Moscow, *Vechernyaa Moskva*, in Russian, August 31, 2009, *FBIS SOV*, August 31, 2009.

Main Military Prosecutor's office revealed about 3,000 violations costing the state another 380 million rubles, leading a commentator to observe that some these criminal schemes were notable not just for their scope but for their brazenness, "one gets the impression that these persons were not afraid of anything."[27]

Therefore it is no surprise that the armed forces are not receiving modern weapons (although corruption is not the only reason for this failure). Another recent audit revealed that "At present the share of the modern types of weapons and hardware that are supplied to the Russian army and navy is not more than 6 percent." And the situation in the high-tech sectors of the military: ships, missiles, and space hardware is especially difficult.[28] Insofar as the defense reform's ultimate success is predicated on the effective production and distribution through the armed forces of modern weapons, this failure jeopardizes the defense reform.

Likewise, in the Ministry of Interior Minister Rashid Nurgaliyev gave regional law enforcement chiefs a month in 2009 to clear out the corruption in their midst or be sacked for failure to control their units or because they too are implicated in the corruption. Nurgaliyev revealed that in the first six months of 2009, a total of 274 criminal proceedings have been launched against Ministry Chiefs at various levels, leading in some cases to outright dismissals. In addition the investigation uncovered 44,000 violations by law enforcement officials, involving 2,500 crimes committed by law enforcement agency employees.[29]

That aforementioned attitude of not fearing anything exemplifies the scope of the problem despite the current pressure to uncover such cases. And the corruption of the government as a whole in Russia needs no explication here given Medvedev's ongoing anti-corruption campaign. Indeed, these cases show that some sense of the scope of this criminality is now becoming public as part of Medvedev's campaign which has been reinvigorated insofar as the military is concerned. Arguably Medvedev's failure to date to uproot this pervasive criminality is what has led to the recent disclosures of corruption in numerous sectors of state and military activity. For example, in the military recent figures show that the number of crimes committed by the military during 2008 rose by nine percent and the crime rate in the military was the highest among the security related agencies in Russia (this is what is in the report, and given the notorious corruption of the police this is a frightening claim). Military prosecutors completed investigations of 12,000 crimes and brought 80 percent of cases to court, including 12 cases against high-ranking military officers.[30] And in the first half of 2009, military investigators completed proceedings of 6,296 crimes, almost 10 percent

---

27 Yuri Gavrilov, "Robbery to Order," Moscow, *Rossiyskaya Gazeta*, in Russian, June 17, 2009, *FBIS SOV*, June 17, 2009.

28 Moscow, *ITAR-TASS*, in Russian, September 23, 2009, *FBIS SOV*, September 23, 2009.

29 Moscow, *Vesti TV*, in Russian August 30, 2009, *Open Source Center, Foreign Broadcast Information Service Central Eurasia* (Henceforth *FBIS SOV*), August 30, 2009.

30 "Crime Rate In Russian Military Rises 9% in 2008," *RIA Novosti*, March 26, 2009.

more than in 2008 while there are also reports of falling crime rates in the Ministry of Emergency Situations and the Ministry of Interior.[31] Nevertheless the number of cases in this sector involving the abuse of authority for "mercenary" reason is increasing, as is the overall military crime rate.[32] Thus, "The scope of corruption in the military has not been decreasing, while bribe-taking has been on the rise, said Col. Konstantin Belyayev of the Main Military Prosecutor's Office. Last year, more than 2,400 corrupt deals were uncovered. The incidence of fraud increased almost 1.5 times, and bribe-taking and abuse of office became more common."[33]

Yet already in July 2009, the Chief Military Prosecutor announced that crimes committed by officers had reached "unprecedented levels." During 2008, officers had committed 4,159 crimes, including 1,754 corruption-related offenses, a 38 percent increase over 2007. Meanwhile already by June 2009 they had committed over 2,000 crimes, or one in four of total crimes, an increase of seven percent on a year-on-year basis. While many of these crimes involve physical assaults on service personnel (over 5,430 personnel reporting such assaults); one third of the crimes involved corruption. Since 2004 the number of Russian generals and admirals prosecuted for corruption had increased by almost seven times.[34] Official figures calculate that these cases of corruption resulted in losses of at least 2.2 billion rubles ($78.6 million) to the state budget in 2008.[35] Finally the evidence of the military forces and its leadership's collusion with organized crime is also now coming to light. The U.S. Cyber Consequences Unit recently reported to the U.S. government that:

> Denial of service and Web defacement attacks launched in 2008 against Georgian websites were carried out by Russian civilians and sympathizers rather than the government but were coordinated with the invasion of the former Soviet state and had the cooperation of both the Russian Army and organized crime, according to a report being released today to U.S. government officials.[36]

This connection is not surprising given extensive reporting of the links between major energy firms like the notorious Rosukrenergo, a key middleman in Russo-

---

31   "Interview With Lt. General A.S. Surochkin, Deputy Chairman of the Investigations Committee Attached to Russian Federation Prosecutor's office and Head of the Military Investigations Directorate by Aleksandr' Kots, Moscow," *Komsomolskaya Pravda*, in Russian, September 7, 2009, *FBIS SOV*, September 7, 2009.

32   Ibid.

33   Simon Saradzhyan, "Russia in Review," citing *Interfax*, February 10, 2011.

34   "Crime Among Russian Military Officers Highest in Decade," *RIA Novosti*, July 9, 2009.

35   Ibid.; "Crime Soars to 10-Year High in Army," *Moscow Times*, July 10, 2009.

36   William Jackson, "Russian Military, Organized Crime In On Cyberattacks Against Georgia," *Government Computer News*, August 17, 2009.

Ukrainian gas deals and leading figures of Russian organized crime and analogous links throughout Eastern Europe.[37]

Since cases that are uncovered are only a fraction of the sum total of criminal activity in any organized social environment, it become clear that we are witnessing the overall degradation of the Russian military and government. It is not too much to say as do many European and U.S. governmental analysts and officials, that we see a criminal, if not Mafia, state (their term).[38] Indeed, no military organization is so isolated from the state and society that its degradation does not both imply and rebound back upon the overall degeneration of that state and society. For example, recent investigations have uncovered figures that were shocking even to the Russian government concerning the brutality and venality of the police forces and the level of criminal violations among them. According to the available statistics, the law enforcement [agencies] are far ahead of the other corruption-prone bodies of power. In 2008 a total of 3,329 police were punished for bribes, in contrast to 433 employees in the health service and 378 in education. According to police, 2,516 crimes committed by police and federal migration service personnel have been identified in January-July, including 1,600 cases of abuse of office.[39]

This last charge that amounts to the criminalization of the state is not as surprising as it may seem, for Russian and foreign observers have long pointed to the integration of criminal elements with both the energy, intelligence and defense industrial sectors of the economy and as an instrument of Russian foreign

---

37   Moscow, *Gazeta.ru*, in Russian, July 27, 2009, *FBIS SOV*, July 27, 2009; Author's conversations with members of European foreign ministries and intelligence services, 2008; Keith C. Smith, *Russian Energy Politics in the Baltics, Poland, and the Ukraine: A New Stealth Imperialism?* (Washington, D.C.: Center for Strategic and International Studies, 2004); Anita Orban, *Power, Energy, and the New Russian Imperialism* (Washington, D.C.: Praeger, 2008); Edward Lucas, *The New Cold War: Putin's Russia and the Threat to the West* (London: Palgrave Macmillan, 2008); Robert Larsson, *Nord Stream, Sweden and Baltic Sea Security* (Stockholm: Swedish Defense Research Agency, 2007); Robert Larsson; *Russia's Energy Policy: Security Dimensions and Russia's Reliability as an Energy Supplier* (Stockholm: Swedish Defense Research Agency, 2006); Janusz Bugajski, *Cold Peace: Russia's New Imperialism* (Washington, D.C.: Center for Strategic and International Studies, Praeger, 2004) passim; Richard Krickus, *Iron Troikas* (Carlisle Barracks, PA: Strategic Studies Institute of the U.S. Army War College, 2006); Valery Ratchev, "Bulgaria and the Future of European Security," paper presented to the SSI-ROA Conference, "Eurasian Security in the Era of NATO Enlargement," Prague, August 4-5, 1997; Laszlo Valki, "Hungary and the Future of European Security," ibid.; Stefan Pavlov, "Bulgaria in a Vise," *Bulletin of the Atomic Scientists* (January-February 1998): 28-31; Moscow, *Izvestiya*, in Russian, June 19, 1997, in *FBIS SOV*: 97-169, June 18, 1997; Sofia, *Novinar*, in Bulgarian, April 10, 1998, in *Foreign Broadcast Information Service, Eastern Europe* (hereafter *FBIS EEU*): 98-100, April 13, 1998.

38   Author's conversations with European foreign ministry and intelligence officials, 2008.

39   "Courts Take unduly Soft Line Toward Police Violence-Media," *ITAR-TASS*, August 17, 2009.

policy in Eastern Europe.[40] Accordingly, summarizing a great deal of evidence, Janusz Bugajski observes that such criminal penetration of Central and Eastern Europe, including the members of the CIS, is a major security concern to those governments because these criminal networks both destabilize their host countries and render services to political interests in Moscow.

> The Russian *Mafiya* greatly expanded its activities throughout the region during the 1990s and established regional networks in such illicit endeavors as drug smuggling, money laundering, international prostitution, and migrant trafficking. In some countries, Russian syndicates have been in competition with local gangs, while in others they have collaborated and complemented each other. Analysts in the region contended that Russian intelligence services coordinated several criminal groups abroad and directed a proportion of their resources to exert economic and political influence in parts of Eastern Europe.[41]

---

40   Vitaly Shlykov, "The Economics of Defense in Russia and the Legacy of Structural Militarization," in *The Russian Military: Power and Purpose*, edited by Steven E. Miller and Dmitri Trenin (Cambridge, MA: MIT Press, 2004), pp. 160-182; Vitaly Shlykov, "The Anti-Oligarchy Campaign and its Implications for Russia's Security," *European Security*, XVII, 2, 2004: 11-128; Leonid Kosals, "Criminal Influence/Criminal Control over the Russian Military-Industrial Complex in the Context of Global Security," *NATO Defense College Research Paper* (March 1, 2004), pp. 6-8; Moscow, *Ekho Moskvy in Russian*, June 4, 2004, *FBIS SOV*, June 4, 2004, *Moscow, ITAR-TASS*, April 14, 2005, *FBIS SOV April 14, 2005; Moscow Center TV* in Russian, September 30, 2003; *FBIS SOV*, October 1, 2003; Moscow, *Moskovskaya Pravda*, in Russian, April 17, 2003, FBIS SOV, April 17, 2003; Moscow, "Interview With OAO Gipromez General director Vitaly Rogozhin," *Rossiyskaya Gazeta*, in Russian, July 13, 2005, *FBIS SOV*, July 13, 2005; Janusz Bugajski, *Cold Peace: Russia's New Imperialism* (Washington, D.C. and Westport CT: Praeger Publishers, 2005); Richard J. Krickus, "The Presidential Crisis in Lithuania: Its Roots and the Russian Factor," Remarks at the Woodrow Wilson Center, Washington, D.C., January 28, 2004, provided by the kind consent of Dr. Krickus; Richard Krickus, *Iron Troikas* (Carlisle Barracks, PA: Strategic Studies Institute, U.S. Army War College, 2006); Keith C. Smith, *Russian Energy Politics in the Baltics, Poland, and the Ukraine: A New Stealth Imperialism?* (Washington, D.C.: Center for Strategic and International Studies, 2004); Tor Bukevoll, "Putin's Strategic Partnership With the West: the Domestic Politics of Russian Foreign Policy, *Comparative Strategy* XXII, 3, 2003: 231-3; Stefan Pavlov, "Bulgaria in a Vise," *The Bulletin of the Atomic Scientists* (January-February, 1998): 30, Robert D. Kaplan, "Hoods Against Democrats," *Atlantic Monthly* (December, 1998): 32-6. As Foreign Minister Igor Ivanov said "Fuel and energy industries in the Balkans are totally dependent on Russia. They have no alternative." "Ivanov on Foreign Policy's Evolution, Goals," *Current Digest of the Post-Soviet Press* (*Henceforth* CDPP), L, 43 (November 25, 1998): p. 13; U.S.-Slovakia Action Commission: Security and Foreign Policy Working Group: Center for Strategic and International Studies, and Slovak Foreign Policy Association, *Slovakia's Security and Foreign Policy Strategy*, 2001, Czech Security Information Service, *Annual Report 2000*.
41   Janusz Bugajski, *Back to the Front: Russian Interests in the New Eastern Europe*, Donald Treadgold Papers, 41 (Seattle: University of Washington Press, 2004), p. 25.

Bugajski's observations correspond to the findings of many other researchers and East European officials concerning the linkages among business, state, intelligence, and organized crime. Thus it has long been known that throughout Eastern Europe and the CIS that the Russian state, intelligence services, energy firms, and organized crime, all collaborate together on behalf of Russian interests. As the record shows they seek to gain access to legitimate business firms, control key sectors of the economy, media, subvert political parties and buy political influence and politicians throughout the region.[42]

Because of the fact that, as Dmitry Trenin has remarked, "Russia is governed by the people who own it," office and property, as in medieval times and in the Soviet Nomenklatura, are one and the same. Power leads to wealth and property and vice versa. Indeed, it cannot be otherwise in such a system. And since this is a system that has systematically freed the executive from any accountability to the media, parliament or anyone else and therefore lacks a concept of the rule of law or of the sanctity of contracts and private property, this outcome is hardly surprising. Recent reports in the Russian press give some indication of the scope of the problem. Medvedev himself has announced what everyone knew, namely that official positions are bought and sold.[43]

Thus this criminality is not confined to the armed forces or security sector but rather it epitomizes the way in which governing occurs throughout the state. Indeed Dmitri Simes and Paul Saunders recently called it the glue that holds together the

---

42 Keith C. Smith, *Russian Energy Politics in the Baltics, Poland, and the Ukraine: A New Stealth Imperialism?* (Washington, D.C.: Center for Strategic and International Studies, 2004); Author's conversations with members of European foreign ministries and intelligence services, 2008; Anita Orban, *Power, Energy, and the New Russian Imperialism*, (Washington, D.C.: Praeger, 2008); Edward Lucas, *The New Cold War: Putin's Russia and the Threat to the West* (London: Palgrave Macmillan, 2008); Robert Larsson, *Nord Stream, Sweden and Baltic Sea Security* (Stockholm: Swedish Defense Research Agency, 2007); Robert Larsson, *Russia's Energy Policy: Security Dimensions and Russia's Reliability as an Energy Supplier* (Stockholm: Swedish Defense Research Agency, 2006); Janusz Bugajski, *Cold Peace: Russia's New Imperialism* (Washington, D.C.: Center for Strategic and International Studies, Praeger, 2004), passim; Richard Krickus, *Iron Troikas* (Carlisle Barracks, PA: Strategic Studies Institute of the U.S. Army War College, 2006); Valery Ratchev, "Bulgaria and the Future of European Security," paper presented to the SSI-ROA Conference, "Eurasian Security in the Era of NATO Enlargement," Prague, August 4-5, 1997; Laszlo Valki, "Hungary and the Future of European Security," ibid.; Stefan Pavlov, "Bulgaria in a Vise," *Bulletin of the Atomic Scientists* (January-February 1998): 28-31; Moscow, *Izvestiya*, in Russian, June 19, 1997, in *FBIS SOV*, 97-169, June 18, 1997; Sofia, *Novinar*, in Bulgarian, April 10, 1998, in *Foreign Broadcast Information Service, Eastern Europe* (hereafter *FBIS EEU*), 98-100, April 13, 1998.

43 Anton Orekh, "Famine," Moscow, *Yezhenedevnyi Zhurnal Internet Version*, in Russian, July 29, 2008, *FBIS SOV*, July 29, 2008.

disparate groups that constitute Russia's governing elite.[44] But beyond that Simes, Saunders, and Russian analysts alike point out that this pervasive corruption not only impedes foreign and domestic investment, it solidifies a dysfunctional political system where the elite has little genuine concern for the national interest or capability to formulate it and is instead busy feathering its own nest.[45] As Vyacheslav Glazychev writes:

> The way Russia is now run clearly reflects Putin's personality and management style. First, there is a linear scheme of administration, based on the idea of the "vertical," rather than a rule-applying bureaucracy. Second, mutual loyalty forms the basis for selecting one's "team" and is combines with open contempt for the government structure itself. Third, the principle of unilateral command from above combines eclectically with some elements of economic liberalism.[46]

The behavior described here perfectly conforms to this depiction of the current governing reality in both the defense sector and throughout the state as a whole. To overcome this sign of pervasive anomic behavior among the security services and sector we must also overcome it in the state, a tall and dubious order. But it is obvious that the continuation of such trends can only further enfeeble the central government's ability to modernize Russia let alone reform or democratize it. Moreover, it is fraught with dangerous implications for Russia, Russia's internal, if not external security, and its armed forces.

First, this widespread criminality provides powerful disincentives to reforming the conditions that make soldiers the easy prey of veterans and officers. And the uprooting of such phenomena as *dedovshchina* (hazing), enserfment of soldiers, theft, and violence against them by superior officers and veterans is essential to any successful defense reform, which, after all aims at creating a so-called professional army. Despite the reforms to date it is still clear that these phenomena remain and pose a serious problem within the armed forces. Indeed, the reforms during the first six months of 2009 did not lead to a reduction in the incidence of crime or corruption within the armed forces; if anything these manifestations increased.[47] This is not only a question of crime and corruption but of hazing, and violence, including torture against soldiers by officers, suicides, and other non-combat deaths.[48] Military spokesmen suggest that this problem may continue because even as the officer corps is downgraded, those remaining are not trained

---

44 Dmitri K. Simes and Paul J. Saunders, "The Kremlin Begs To Differ," *The National Interest*, 104 (November-December, 2009): 39.

45 Ibid., 38-42; Ivan Krastev, Mark Leonard and Andrew Wilson (eds), *What Does Russia Think?*, European Council on Foreign Relations, 2009, www.ecfr.eu, pp. 1-52.

46 Vyacheslav Glazychev, "The 'Putin Consensus' Explained," ibid., p. 13.

47 Roger N. McDermott, "Crisis Looms In Russia's Armed Forces," *Asia Times*, September 4, 2009.

48 Ibid.

or equipped to deal with a new army and others may resist losing their perquisites. Worse yet in 2009 figures suggest that not only is the Russian army drafting people with a criminal record for the first time in this decade, but that their number amounts to more than half of those drafted since fall 2008.[49] While the government is now introducing chaplains for the armed forces to introduce some form of moral counseling and attempting other procedural reforms to stop this trend, if the new army remains a home for criminals and brutes that will defeat the entire purpose of the reform.

More grandly, this widespread brutality and corruption lead the military leadership, much of which directly benefits from this state of affairs, to resist reforms and create powerful obstacles to reforms that would lead to a genuinely modern, and truly professional army where soldiers have enforceable legal rights and recourse against accountable colleagues and officers rather than perpetuate the continuing treatment of enlisted men as serfs and "baptized property" (the term coined by the nineteenth century dissident (Alexander Herzen to describe serfs). Moscow's earlier inability and refusal to reform its military, end conscription, and institute a genuinely professional military leads to an armed force composed of the uneducated, physically, morally, and mentally unfit, and widespread brutality and corruption which militates against an army that can, except for certain specialized forces, effectively use high-tech weaponry. Under the circumstances it is not surprising that Chief of Staff, General Nikolai Makarov admitted in 2008 that the army was not ready for twenty-first century warfare.[50] And this was hardly the only set of reasons why the army was so backward compared to contemporary requirements.

Certainly the pervasiveness of these pathologies precludes creation of a truly professional army in any sense of the word. This is not merely a question of men and women being paid well for their services to the state, nation, and military. It also is a question of inculcating in the armed forces the sense of professionalism, of belonging to a profession with a genuine ethic of patriotic service. This ethic arguably is that of a profession not that of a bureaucracy although in Russia's case, while we have the pathologies of bureaucratic procedure and an immense state, we certainly do not even have a bureaucracy in the sense of a disinterested and nonpartisan corps of public servants. As a result the whole notion that commanding officers can lead the armed forces in such a way as to inculcate this professionalism and an ethic of it among the men under their command flies out the window. Instead we have an army like the one seen in Georgia and described above by Makarov.[51]

---

49   Ibid.

50   Pavel Felgenhauer, "Russia's Radical Military Reform in Progress," *Eurasia Daily Monitor*, November 20, 2008.

51   See Don M. Snider, Paul Oh and Kevin Toner, *The Army's Professional Military Ethic in an Era of Persistent Conflict* (Carlisle Barracks, PA: Strategic Studies Institute, U.S. Army War College, 2009), p. 2.

Yet at the same time the reform has paradoxically given a new impetus to corruption and criminality within the armed forces that may help explain the rise of such incidents even as the reform is occurring. Marc Galeotti offers the following reasons for the new impetus towards corruption. The reform takes place in a context of constantly rising defense appropriations, including for 2010. Much of this will go to the reform, specifically raising salaries and professionalization, that is, the "recruitment" of "professional" soldiers at higher rates of pay and improved conditions and housing. Already some officers receive bonuses that triple or quadruple their basic pay. Consequently officers are scrambling for bonuses and to avoid dismissal as the armed forces downsize.[52]

> This has created massive opportunities for corruption. Senior officers and those within the personnel directorates can demand and expect substantial bribes for their recommendations. According to some Defense Ministry sources, the going rate can be the equivalent of a full year's salary in return for guaranteeing continued employment on the higher pay scale. Furthermore, the Defense Ministry is gearing up for a massive campaign of refurbishing and replacing rundown barracks and other facilities. This opens up opportunities for a wide range of money-making ventures from selling off second-hand furniture and equipment (which is then logged as having been destroyed) to manipulating bidding by contractors to secure government contracts.[53]

And the continuing insurgencies in the North Caucasus contribute greatly to this state of affairs.

> If crimes by officers throughout the country in general hold to their normal level, meaning that every fourth criminal is an officer, then, in the 42nd Motorized Rifle Division, which deployed to Chechnya, the situation is much worse, with more than half the crimes in the unit committed by the officer corps. The situation is also bad in the Airborne Troops, the Space Troops, the Air Force, the Volga-Urals Military District, North Caucasus Military District and the Moscow garrison. There almost a third of all crimes reported last year was committed by officers.[54]

Crime has been not limited to lower and mid-level officers. The same source noted that "In 2004, only three generals were tried, but in 2008, 20 were." The bottom line is that officer crimes are out of control. "The crime rates are the highest over

---

52 Marc Galeotti, "Have Russia's Dirty Generals turned on Shamanov?," *Radio Free Europe Radio Liberty*, September 29, 2009.

53 Ibid.

54 "Military Honor is Being Disbanded," *Moskovskiy Komsomolets*, July 30, 2009, http://www.mk.ru/, also in WNC, August 1, 2009.

the past ten years. Officers are responsible for more than 2,000 crimes with one-third of these linked to corruption."[55]

Thus pervasive corruption, criminality, and brutality have become major causes of Russia's inability to deal effectively with the mounting threats in the North Caucasus as they are helping to turn the local population away from Moscow and to the Islamic fundamentalists who are leading the revolts against Russian rule. Any objective account of this insurgency cannot overlook the seriousness of the Islamic threat and its equal levels of violence and terror against the population.[56] Nevertheless the fact remains that every fifth incident in the Army involves servicemen from the North Caucasus where indeed there is no problem recruiting soldiers. Obviously the potential material rewards of service plus the martial traditions of the region are attractive to local men, especially as this remains the poorest region of Russia. Indeed, men are now bribing recruiters to get into the service in an ironic reversal of past practice that involved bribes to be exempted.[57] These soldiers bring their own culture and a propensity for ethnic organization of parallel discipline structures into the army leading to violence, discipline problems, inter-ethnic conflicts within units, and of course, criminality. But if the Russian army continues to experiment with ethnic based units it runs the risk of intensifying the problems already discerned here.[58]

In the Caucasus numerous reports show that Russian armed forces (it is unclear if this is the regular army or the VVMVD) operate as death squads. This is a long-standing phenomenon growing out of the long war in Chechnya. It also appears to be pervasive throughout the North Caucasus. Such operations are particularly centered in the units of the various Special Forces operating there and are looked on with a blind eye by higher authorities.[59] They kidnap and kill people with seeming impunity and Russian human rights organizations suggest that there exists a correlation between this violence and the rising tide of insurgency in both Chechnya in particular and the North Caucasus as a whole.[60] Memorial's (The organization established to preserve a living memory of Stalin's crimes) Aleksandr' Cheraskov observed that these death squads target young men of military age which only makes them more susceptible to recruitment by rebel groups.[61] And Ludmilla Alexeyeva, the head of the Moscow Helsinki Group said:

---

55   "Crime Rates in Army Highest Over Past Ten Years—Prosecutor," *ITAR-TASS*, July 9, 2009, also in WNC, July 10, 2009.

56   Gordon M. Hahn, *Russia's Islamic Threat* (New Haven, CT and London: Yale University Press, 2007).

57   Aleksandr' Stepanov, "Mountain Law," Moscow, *Nasha Versiya*, in Russian, August 31, 2009, *FBIS SOV*, August 31, 2009.

58   Ibid.

59   Mark Franchetti, "Russian Death Squads 'Pulverize' Chechens," *Times Online*, April 26, 2009, www.timesonline.co.uk.

60   "Russia Running Death Squads in the Caucasus," *The News*, September 3, 2009, www.thenews.com.pk.

61   Ibid.

> What we see now in all these (Caucasus) republics is a civil war between the
> security forces and the clandestine fighters, and between the security forces and
> the local population—In the end we will lose the North Caucasus. The Russian
> president doesn't wish this, of course, but he has no control over his own security
> forces.[62]

And that is precisely the point. After 10 years of unsparing brutality on all sides
in Chechnya, it and its neighbors are aflame with no end to these conflicts in
sight. Much of this is directly traceable to the violence and corruption of the
Russian armed forces that continually undermines the real security of the Russian
Federation and which itself is only a partial manifestation of the larger and even
more endemic corruption and brutality of the government. As Trotsky memorably
wrote, "The army is a copy of society and suffers from all its diseases, usually at
a higher temperature. The trade of war is too austere to get along with fictions and
imitations. The army needs the fresh air of criticism. The commanding staff needs
democratic control." Although he thought the Red Army was democratic we know
better today. And today's Russian army clearly reflects the pathologies of its state
and society. If its relations with society and the state are to improve then the first
and primary precondition for such change is the removal of these pathologies from
the Russian state. Absent that long-term and uphill struggle Russia will not only be
a security threat to its interlocutors, but even more to its own people.

---

62 "Russia Running 'Death Squads' in Caucasus: Rights Groups," *The Dawn*, www.
dawn.com, September 3, 2009.

Chapter 4

Civil-Military Relations and the American Way of War[1]

John Allen Williams

Introduction

The international system is well into an era of conflict where traditional notions of the "American Way of War" are decreasingly relevant. Indeed, an earlier tradition of fighting "small wars" may be more appropriate to the new challenges faced.

The United States military has sustained significant casualties in at least two wars still underway (including Iraq, where conflict will surely continue beyond the expected withdrawal of U.S. combat forces by the end of 2011). From a civil-military relations perspective this period is notable by a partial, yet massive mobilization that involves both active and reserve military components, but not the civilian society except insofar as reserve forces are now heavily involved.

The civilian society both empowers and constrains U.S. military forces, not only in terms of human capital and political support, but in evolving civilian norms relating to the conduct of war and the composition of military forces. Military structure, the definition of victory, and the means that may be employed to achieve it are affected by these civilian perspectives. Conversely, civilian expectations are affected by the perception of civilians of military success, the costs incurred, and the military institution itself. This chapter explores these phenomena and discusses their implications for the U.S. armed forces and for American civil-military relations in the future.

The Changing Environment[2]

One should be cautious in making predictions about strategic futures. Extrapolations from current trends are useful, but not sufficient. They can also be misleading,

---

1 The author thanks Stephen J. Cimbala, Timothy Hazen, Eric Morse, and Michael P. Noonan for their assistance in thinking through and preparing this chapter.

2 See John Allen Williams, "Statement Before the U.S. House Armed Services Committee, Subcommittee on Oversight and Investigations," panel on "Charting the Course for Effective Professional Military Education," September 10, 2009. The subcommittee report is available at U.S. House of Representatives, Committee on Armed Services,

as they do not incorporate any radical discontinuities that may occur. Strategic surprises are by their nature surprising, even if it seems from after the fact that they should have been predicted. (Pearl Harbor and even September 11 come to mind.) This is complicated by the complex nature of the evolving security landscape. As defense analyst John Collins pointed out, "Modern military strategists ply their trade in volatile environments that are fraught with more uncertainties, complexities, and ambiguities than Clausewitz imagined."[3]

The International Environment

A project completed some years ago discussed the international threat through three eras: before, during, and after the Cold War and analyzed how the evolution of the threat affected areas of civil-military relations ranging from force structure to major mission definition to the dominant military professional to public attitudes toward the military and to issues of gender and sexual orientation. There appeared to be many similarities among Western democracies in the way in which these variables evolved. The final period was called the "Postmodern" era, marked by an increased importance of threats of ethnic violence and terrorism emerging from within nation-states.[4]

Through most of American history even the most dangerous security challenges were fairly straightforward and yielded to primarily military solutions. Although there was no lack of politics in determining what the solutions should be, with the possible exceptions of the Philippine insurrection and the Vietnam War, military operations could be conducted with minimal concern on the part of the military about the views of the enemy civilian population.[5]

After the September 11, 2001 attacks, the model was expanded in view of the emergence of transnational non-state actors that clearly now posed a significant threat. It seemed that a new era was at hand, named the "Hybrid" era.[6] This era

---

Subcommittee on Oversight and Investigations, "Another Crossroads? Professional Military Education Two Decades After the Goldwater-Nichols Act and the Skelton Panel," April 2010, http://democrats.armedservices.house.gov/index.cfm/files/serve?File_id=d4748d4a-b358-49d7-8c9a-aa0ba6f581a6/

3   John M. Collins, *Military Strategy: Principles, Practices, and Historical Perspectives* (Washington, D.C.: Brassey's Inc., 2002), p. 9. This sophisticated introduction to military strategy deserves a wide audience.

4   Charles C. Moskos, John Allen Williams, and David R. Segal, *The Postmodern Military: Armed Forces After the Cold War* (New York: Oxford University Press, 2000).

5   On the Philippine Insurrection and the U.S. Army's response to it, see Sam C. Sarkesian, *America's Forgotten Wars: The Counterrevolutionary Past and Lessons for the Future* (Westport, CT: Greenwood Press, 1984).

6   John Allen Williams, "The Military and Society Beyond the Postmodern Era," *Orbis: A Journal of World Affairs*, 52:2 (Spring 2008): 199-216. For a discussion of hybrid wars, a notion closely related to the hybrid era, see Frank G. Hoffman, *Conflict in the 21st Century: The*

includes international, transnational, and subnational threats. These complicated the challenges faced by the military, which must now contend with a full spectrum of operations. Marine Corps General Charles Krulak captured this notion well with his concept of the "three block war," in which high-intensity, peacekeeping, and humanitarian operations occur simultaneously in close proximity to one another.[7]

The Domestic Environment

The domestic environment is also changing, whether as a result of the changing international environment or other social forces. First, society has become more accepting of diversity of all kinds, whether of race, religion, gender roles, or sexual orientation. We have not yet arrived at a point where individuals suspend moral judgment about the behavior of others, but the range of behavior considered acceptable or even normal is much greater than it was even a dozen years ago. While some may find this development appalling, it means that military service is now open to a far broader range of people than before and inside the military opportunities are increasing for women, in particular. This is not due to a sudden epiphany on the importance of fairness, but rather a realization that such discrimination makes little sense morally or practically. Indeed, the case can be made that racial integration in the U.S. military did not happen with President Truman's Executive Order to that effect, but rather because of personnel needs for the Korean War.

Second, the United States is no longer in denial about the existence of genuine threats to its citizens. While the nuclear terror of the Cold War is over, there are new threats that pose a real danger. The 1990s were in many respects a vacation from reality because while the old danger had gone away, most Americans failed to imagine and prepare for the new ones. Reasonable people differ as to what the response to these should be, but the reality of the threat should not be ignored.

Third, we have entered into an era of fiscal austerity unparalleled in recent history. The near meltdown of the American economy and the slow recovery from the "great recession" have convinced the public that something must be done. As in the case of the proper responses to security threats, there is little consensus about who should be expected to sacrifice what, but fiscal retrenchment is clearly on the table politically. It is easy to be intolerant of waste, fraud, and abuse, but ultimately economic considerations will result in cutting not only fat but what many will consider bone from the military. The luxury to prepare for everything

---

Rise of Hybrid Wars (Arlington, VA: Potomac Institute for Policy Studies, December, 2007), http://www.potomacinstitute.org/index.php?option=com_content&view=article&id=77:-conflict-in-the-21st-century-the-rise-of-hybrid-wars&catid=40:books&Itemid=62/

7   Gen. Charles C. Krulak, "The Strategic Corporal: Leadership in the Three Block War," *Marines Magazine* (January 1999), http://www.au.af.mil/au/awc/awcgate/usmc/strategic_corporal.htm. Accessed September 5, 2009.

in the hope that something will be relevant in the event of need can no longer be indulged.

Fourth, civilian society does not give special dispensation to the military to behave in ways unacceptable in a civilian context, with the exception of actual combat – and even in that situation soldiers are held to a high standard. Civilian mores dominate, and military members are expected to observe them. This trend may have been developing for some time, but it was clear after the "Tailhook" scandal,[8] that it had arrived.

Finally, the military does not wish to operate domestically, except perhaps for flood relief or humanitarian operations of other sorts. Military resources are often the only ones available quickly in the event of some natural or human-caused catastrophe, but the military prefers to operate in support of other agencies, such as the Federal Emergency Management Agency (FEMA). On its part, society has long been wary of domestic intervention by the military. The *posse comitatus* restrictions have been in place since 1878, and by and large have been strictly observed. However, these restrictions may need to be waived in view of some terrorist threat or in the aftermath of a civil emergency. It may be that the military is the only force available to restore order or – and this is a very troublesome consideration – to enforce a quarantine.[9] The American people have a very high opinion of the military, and this is reinforced by the perception that the military is operating somewhere else rather than here at home. Unpopular domestic operations will severely erode military popularity and prestige.

The American Way of War

Discussions of the American way of war appropriately begin with the great military historian Russell F. Weigley. For Weigley, the important distinction is between a *strategy of attrition*, which is an indirect strategy designed to wear down opponents over time, and a *strategy of annihilation*, which is designed to defeat the military forces of an enemy. Over time, Americans developed a preference for the latter:

> In the history of American strategy, the direction taken by the American conception of war made most American strategists, through most of the time span of American history, strategists of annihilation. At the beginning, when American military resources were still slight, America made a promising beginning in the nurture of strategists of attrition; but the wealth of the country

---

8   The best discussion of this incident and the fallout from it is William H. McMichael, with a foreword by Charles C. Moskos, *The Mother of All Hooks: The Story of the U.S. Navy's Tailhook Scandal* (Edison, NJ: Transaction Publishers, 1997). The impact of this event and its social and political repercussions cannot be overstated.

9   Williams, "The Military and Society Beyond the Postmodern Era," p. 205.

and its adoption of unlimited gains in war cut that development short, until the strategy of annihilation became characteristically the American way of war.[10]

Colin Gray offers a more expansive list of the characteristics of the American way of war. There is not space here to go into the details of these characteristics, but the reader will get a notion of the factors Gray believes are most important in understanding how Americans fight. He describes 13 attributes, saying that the American way of war is apolitical, astrategic, ahistorical, problem-solving and optimistic, culturally challenged, technology dependent, focused on firepower, large-scale, aggressive and offensive, profoundly regular, impatient, logistically excellent, and highly sensitive to casualties.[11] Notice that these characteristics are those of a large and wealthy country fighting regular forces. As the strategic landscape changes and irregular forces assume more importance, these characteristics can become a liability.

Antulio Echevarria concurs that the American military is highly focused on combating enemy forces, as opposed to other emphases that would be more useful in the current environment. He noted in 2004, before the emphasis on counterinsurgency regained its priority, that "the current American way of war focuses principally on defeating the enemy in battle. Its underlying concepts ... center on "taking down" an opponent quickly, rather than finding ways to apply military force in the pursuit of broader political aims."[12] Of course, the U.S. is most comfortable fighting a war in which "[t]he enemy must be the state and not an insurgency, and we need to march on the adversaries capital and topple the government."[13]

Many discussions of the American way of war focus on the American reliance on high technology for its security. One advantage of this approach for a technologically sophisticated country such as the United States is that it could use its technological prowess to make up for the relatively smaller size of its military compared to the Soviet Union. This was noted by Michael Melillo, who pointed out that this tendency continued through the discussions about the "Revolution in Military Affairs" and "Network-centric warfare" that would supposedly lift the fog of battle.[14]

---

10   Russell F. Weigley, *The American Way of War: A History of United States Military Strategy* (New York: Macmillan Publishing Co., 1973), p. xxii.

11   Colin Gray, *Irregular Enemies and the Essence of Strategy: Can the American Way of War Adapt?* (Carlisle, PA: U.S. Army War College Strategic Studies Institute, 2006), p. 30, http://www.strategicstudiesinstitute.army.mil/pubs/display.cfm?pubID=650/

12   Antulio Joseph Echevarria, *Toward an American Way of War* (Carlisle, PA: U.S. Army War College Strategic Studies Institute, 2004), p.16.

13   Dominic Tierney, *How We Fight: Crusades, Quagmires and the American Way of War* (New York: Little, Brown and Company, 2010), p. 7.

14   Michael R. Melilo, "Outfitting a Big-War Military with Small-War Capabilities," *Parameters,* 36:3 (2006): 24-5, http://www.dtic.mil/cgibin/GetTRDoc?AD=ADA283762& Location=U2&doc=GetTRDoc.pdf/

Another characteristic is to regard wars as moral engagements of good versus evil. This is useful to generate support for the sacrifices involved, but the problem with such a characterization, as Thomas Mahnken observed, is that it makes difficult the pursuit of limited aims in the policy of something other than complete victory.[15] Mahnken concurs with Gray that there are a number of additional factors, including direct strategies designed to defeat the enemy as quickly as possible, an industrial approach to war, and an emphasis on firepower and technology. He also raises the possibility of an American reluctance to incur battlefield casualties. If so, this tendency may be more apparent among civilian and military leaders than among the general public. It is also widely believed among America's adversaries.[16]

Frank Hoffman emphasizes that "the American Way of War is built around a strategy that employs the vast economic and technology base of the U.S. to grind down opponents with firepower and mass."[17] After Operation Desert Storm, however, dominant strategies no longer focused on total victory, but the military engagements themselves still relied on overwhelming force.[18] Hoffman also points out that traditional expectations of a clear line of demarcation between civilian and military responsibilities are insufficient, since strategic success requires an understanding that these responsibilities overlap.[19]

Hoffman's discussion of this strategic evolution points out another perspective, sometimes described as an alternative way of war. That it certainly is, but it is also fully part of the American military experience. Max Boot said it well:

> Some historians even speak of an "American way of war": war that annihilates the enemy; or that relies on advanced technology and massive firepower to minimize casualties among U.S. forces ... But this is only one way of American

---

15 Thomas G. Mahnken, *United States Strategic Culture* (Washington, D.C.: SAIC, November 13, 2006), p. 9.

16 Mahnken, pp. 10-14. On the casualty issue, Mahnken refers to Peter D. Feaver and Christopher Gelpi, "How Many Deaths are Acceptable: A Surprising Answer," *The Washington Post*, November 7, 1999, p. B3.

17 Frank G. Hoffman, *Decisive Force – The New American Way of War?* (Newport, RI: Naval War College, 1994), p. vi. Available at: http://www.dtic.mil/cgi-bin/GetTRDo c?AD=ADA283762&Location=U2&doc=GetTRDoc.pdf. Alternatively, see Frank. G. Hoffman, *Decisive Force – The New American Way of War* (New York: Praeger Publishers, 1996).

18 Hoffman, *Decisive Force*, p. 149

19 Frank G. Hoffman, "Dereliction of Duty *Redux*?: Post-Iraq Civil-Military Relations," *Orbis*, 52:2 (Spring 2008): 224, 228. The classic study by political scientist Samuel P. Huntington, *The Soldier and the State* (Cambridge, MA: Harvard University Press, 1957), suggests that civilian and military responsibilities can be disentangled. Compare the sociologist Morris Janowitz, who demonstrates—indeed, celebrates—the complex interrelationships between the military and society. See Morris Janowitz, *The Professional Soldier: A Social and Political Portrait* (New York: The Free Press, 1960 and 1971).

war. There is another, less celebrated tradition in U.S. military history – a tradition of fighting small wars. Between 1800 and 1934, U.S. Marines staged 180 landings ... These are the non-wars that Kipling called "the savage wars of peace ..."[20]

Such "Small Wars" are nothing new for the United States, and we may now be in the process of rediscovering one branch of our roots. The extraordinarily thorough and perceptive *Small Wars Manual*, produced by the Marine Corps in 1940, notes the frequency of Marine participation in these wars. Marines were operating abroad every year since the Spanish-American war, with two thirds of the Corps deployed outside the continental United States in 1929.[21]

Readers not familiar with this document may be struck by the sophistication of its analysis. Topics of discussion include phases of small wars, the particular importance of a political strategy in small wars, and close attention to the characteristics of the people in the country of intervention. In a phrase that would be in discussions of current counterinsurgency doctrine, it notes, "... A simple display of force may be sufficient to overcome resistance. While curbing the actions of the people, courtesy, friendliness, justice, and firmness should be exhibited." It continues, "in small wars, tolerance, sympathy, and kindness should be the keynote of our relationship the mass of the population." It goes on to say that these characteristics should not preclude such firmness as necessary, however.[22]

The vital importance of political considerations is basic for success in any war, and some agree with the *Small Wars Manual* that this is especially true in small wars. The difficulty the United States has had prevailing in small wars is partly due to inadequate attention to political matters. Indeed, according Carnes Lord, the vital role of political considerations is a "key characteristic" of small wars.[23] He remarks, "the United States has typically approached small wars as war writ small; it needs to understand them rather as politics writ large."[24]

These notions of an alternative American way of war have in common the idea that the purpose of the military is to combat hostile forces and the enemy population are not themselves the object of attack (although there were certainly serious exceptions to this during the Second World War). U.S. Air Force officer and historian John Grenier offers a different view from his reading of history. In fact, the "normal" American way of war involved precisely such attacks on

---

20 Max Boot, *The Savage Wars of Peace: Small Wars and the Rise of American Power* (New York: Basic Books, 2002), p. v-vi.

21 United States Marine Corps, *Small Wars Manual* (1940), p. 2, http://www.au.af. mil/au/awc/awcgate/swm/index.htm/

22 *Small Wars Manual*, pp. 5, 11, and 13. The quotations are from p. 15 and p. 32.

23 Carnes Lord, "The Role of the United States in Small Wars," *Annals of the American Academy of Political and Social Science* (September 1995): 90.

24 Lord, p. 99.

noncombatant populations and any property that supported enemy fighters.[25] He noted that "the first way of war against Indian noncombatants on the frontier continued to define American war making throughout the nineteenth century... ."[26]

In any case, there is wide agreement that traditional Clausewitzian notions of war are appropriate for only some of the challenges confronted by the United States in the new security era of ambiguous challenges and a lack of moral certainty. Choices do not present themselves as black and white, but rather shades of gray. Solutions involve complicated mixtures of civilian and military approaches, and require coordination of all aspects of national power. Civil-military relations moves far beyond deciding whether the military is "controlled" or not to explore the implications of the close interactions required for national security.

Civil-Military Relations

Given the dramatic changes in the strategic landscape discussed above, it is not surprising that the American way of war is evolving to take account of the new strategic challenges. These changes are having a profound effect on the military and on the relations between the military, civilian institutions, and society.

Further, one cannot overestimate the importance of a dramatically reduced level of resources on the military. Such fiscal decisions are made by civilians, who not only provide resources for the military; they shape the force structure by fiscal decisions. In making these decisions they rely on military expertise, but the manner in which the military should express its views is controversial.

Military Leaders

The Postmodern Military study cited above suggested that the dominant military professional evolved in response to the changing threat. It is clear that the role of combat leader remains essential in a military that expects to win wars rather than deploy as armed social workers. There are other important roles, such as manager/technician (especially given the importance of high technology), soldier-statesman, and soldier-scholar. In the new era, the military may need to assume some missions in stability operations that require a law enforcement function. The evolution of the soldier-constable is inevitable, with the caveat that this is a very dangerous role for the military to perform at home.[27]

Increasingly, military leaders and followers alike must be intellectually flexible —able to understand and adapt to ambiguous situations quickly and effectively.

---

25   John Grenier, *The First Way of War: American War Making on the Frontier* (New York: Cambridge University Press, 2005), p. 1.

26   Grenier, p. 221.

27   Williams, "The Military and Society Beyond the Postmodern Era," p. 202.

The full implications of this for military recruitment and professional education are beyond the scope of this discussion, but it does suggest that an emphasis on technology alone will not produce the leaders needed going forward. Humanities and social science studies are more important than is generally appreciated, especially by the technology-intensive Navy and Air Force.

Military Public Expression

There is a continuing debate about the appropriateness of serving military officers expressing their views on matters of public policy affecting the military, including operations.[28] One clear regulation affects commissioned officers: Article 88 of the Uniform Code of Military Justice prohibits the use of "contemptuous words" against certain public officials.[29] (There are fewer restrictions on the public pronouncements of enlisted personnel, as long as they make it clear their comments are personal.)

There is no prohibition against retired personnel making their opinions known on any issue, although partisan political comments could detract from the esteem in which the military is held. Serving officers are appropriately held to a higher standard, however. At the same time, public discussion of military policies by serving officers can contribute to public understanding. It is especially important to testify truthfully to Congress, as did then-U.S. Army Chief of Staff General Eric Shinseki when asked while testifying before the Senate Armed Services Committee how many troops would be needed in Iraq after a successful invasion. His response of "several hundred thousand troops" did not please his civilian superiors, but the general was correct in his analysis and it was appropriate to answer truthfully.[30]

---

28   For an extended discussion of this issue, see Sam C. Sarkesian, "The U.S. Military Must Find Its Voice," *Orbis* (Summer 1998), pp. 423-4. A contrary view is well expressed in Richard H. Kohn, "The Erosion of Civilian Control of the Military in the United States," *Naval War College Review*, 55 (Summer 2002): 8-59.

29   Title 10 U.S. Code, Section 888, Article 88: "Any commissioned officer who uses contemptuous words against the President, the Vice President, Congress, the Secretary of Defense, the Secretary of a military department, the Secretary of Transportation, or the Governor or legislature of any State, Territory, Commonwealth, or possession in which he is on duty or present shall be punished as a court-martial may direct."

30   See Damon Coletta, "Courage in the Service of Virtue: The Case of General Shinseki's Testimony before the Iraq War," *Armed Forces & Society*, 34:1 (Fall 2007): 109-21 and Paul R. Camacho and William Locke Hauser, "Civil-Military Relations – Who Are the Real Principals?" *Armed Forces & Society*, 34:1 (Fall 2007): 122-37 for differing views on this incident.

Inter-rank Relations

The increasing complexity of military duties has implications not only for the selection and promotion of the right people to perform them; it will inevitably affect relations among the ranks. Increasingly professional enlisted forces and the officers placed over them will relate in ways that are more task-oriented and egalitarian. Of course, the command structure will remain, but the hierarchy will flatten. This has long been true in the military medical communities, but there are also elements of this in highly elite special operations forces.

One of the defining elements of a hierarchy is as a means of channeling communications. Personnel now being recruited into the force are increasingly familiar with social networking sites and other means of electronic communication that are anything but hierarchical, and one should not be surprised that junior personnel share their ideas freely with anyone concerned. This can be a positive development for military that wishes to adapt to the changing environment, but it will be problematic for old hands uncomfortable with informality and unofficial communications links.

Civilian Contractors

There are many functions previously performed by military forces that can be done more efficiently by civilian contractors. Even though up-front direct costs may be higher, the military has the benefit of great flexibility in contracting without incurring long-term personnel commitments—especially retirement costs. This flexibility and economy will be critical as the military tries to do more with less in a shifting strategic environment. This is not a new development, of course, but it is the way of the future.

Contracting for support services, such as administrative and logistic support, is not troublesome in principle. More controversial is the role of civilian contractors on the battlefield itself.[31] Credible allegations of the excessive use of force make it essential to clarify criteria for selection, proper missions, rules of engagement, supervision, and ultimately legal accountability. There may well be a place for armed civilians on or near the battlefield, but their status needs to be clarified. The military cannot be large enough to perform all the functions that need to be accomplished, and given the new fiscal realities, the American way of war will need to accommodate some changes in this area.

---

31   For opposing views on this issue, see Robert D. Kaplan, "Outsourcing Conflict," *The Atlantic*, September 2007, http://www.theatlantic.com/doc/200709u/kaplan-blackwater, and Ralph Peters, "Trouble for Hire," *New York Post*, September 30, 2007, http://www.nypost.com/seven/09302007/postopinion/opedcolumnists/trouble_for_hire. htm. See also Deborah D. Avant, "Contracting for Services in U.S. Military Operations," *PS: Political Science and Politics* XL, 3 (July 2007): 457-60.

Relationship with Civilian Agencies

Although details of the attack remain sensitive, it appears that the successful raid on the Pakistan compound of Osama bin Laden in May 2011 was carried out by a combination of CIA and military Special Forces. While many agencies were concerned with locating and capturing him, probably the overall responsibility lay with the CIA. Many were surprised to hear that the chain of command for the operation ran through the Director of the CIA rather than the Secretary of Defense. The attack showed close coordination, not only among military elements (which is less difficult than it used to be), but between the military and civilian agency. General David Petraeus' return to Washington to run the CIA post-retirement is another indication of the degree of closeness between civilian intelligence and the military.

Bin Laden's death is a recent example of the military dealing regularly with other civilian governmental organizations such as the CIA, FEMA, NGOs and other organizations that work abroad to facilitate in attempts of "nation building."

Integration of Reserves

The closer integration of reserve component forces with the active component accelerated in the wake of the Vietnam War, when the Army reorganized to ensure that any significant conflict would have to involve society form of its reserve forces. Stung by criticisms that were the product of a long and bitter military struggle, the Army Chief of Staff at the time, General Creighton Abrams, put vital support functions into the reserves. No longer would the Army bear the brunt of popular opprobrium toward an unpopular war.

As a result, reserve forces were no longer merely supplemental to active forces, designed to be called up only in the face of the gravest emergency; the forces were closely integrated and reserve forces were used on a regular basis for normal operations. When the missions assigned to the Army and Marine Corps in Iraq and Afghanistan overtaxed the capabilities of the active force to handle them, calling up reserves became a regular occurrence. The active forces are now dependent on reserve forces, including the National Guard, for success.

Repetitive and lengthy periods of active duty are very disruptive to reserve personnel, who also wish to maintain their civilian jobs or professions. Reservists' families do not necessarily live near a military base, making the assistance available there more difficult to access. Despite laws prohibiting discrimination in such cases, employers surely consider the likely availability of a worker before assigning him or her to a critical function. In addition, some reservists suffer a loss of income that can harm their financial stability.[32]

---

32 See Bradford Booth, Mady Weshsler Segal, and D. Bruce Bell, *What We Know About Army Families: 2007 Update* (Prepared for the Family and Morale, Welfare, and

General Abrams was correct that the change he implemented would bring the military and society closer together. Quite aside from the effect this may have on popular support for conflicts and the military fighting them, this is a linkage that is useful to reinforce. At a time when military service is not widespread, particularly among societal elites, it serves the useful purpose of making the military more widely understood – not only among the reservists finding out what active service is really like, but among their families and contacts in their communities.

Personnel Issues

Military personnel issues are greatly affected by the changing strategic situation, changing civilian norms, and the evolving American way of war. This is especially apparent when considering issues of gender, sexual orientation, and recruitment.

With respect to gender, very few limitations on the service of women in the U.S. military remain. The significant exceptions to this involve exclusions from close combat. Women may not be assigned directly to infantry, artillery, or armored units or to the Special Forces. Prohibitions against women serving in submarines are falling, and restrictions on women's participation in naval aviation have long since been removed. As societal norms of equality and equal opportunity become stronger, considerations of military effectiveness become less persuasive. Indeed, many argue that there would be no impact on military effectiveness if all combat specialties were open to women. It is beyond the scope of this discussion to resolve this issue, beyond the observation that reasonable people differ on it and the eventual elimination of remaining restrictions is inevitable.

Evolving societal norms with respect to sexual orientation finally resulted in a change in the congressionally-mandated exclusion of open homosexuals in the military. The "Don't Ask, Don't Tell" policy, designed as a workable, if imperfect, compromise, will be eliminated by the time these words are read.[33] The military was given time to ease the transition, but many issues have not been resolved and this is a subject on which feelings run very high. Nevertheless, the military is reflecting society in this change.

Recruitment

Military conscription ended in 1973, although systems remain in place to conduct compulsory call-ups if necessary. The lack of conscription, combined with the relatively small size of the military, means that relatively few citizens will need to

---

Recreation Command by Caliber: an ICF International Company, 2007), p. 65.

33   The author of this policy, military sociologist Charles Moskos, often remarked that "this is a compromise that everybody hates." He regarded that as evidence that it was about right.

serve or even focus on military issues. This has not produced anti-military feelings, however, and quite the opposite is true. The military is among the most admired institutions in the country and military people are routinely stopped in the street and thanked for their service.[34] It may be that a lack of familiarity with the military will cause people to overestimate its capabilities or to believe that there actually can be such a thing as a "surgical strike," but there is a decided lack of animus.

While reflective of societal norms, the military is not representative of society from the perspective of socioeconomic status. Only a genuinely fair system of conscription could accomplish this, and even with such a system in place it is often possible for high status individuals to game the system. It is often described as an "all volunteer" military, but in fact it is not simply spontaneous volunteerism that fills the ranks; the military must provide and advertise inducements to convince young people to join it. From that perspective, it may be more appropriate to regard the force as an "all recruited" military. As noted above, reserve forces also link the military and society, but this is not the best substitute for a genuinely representative active force.

The technologically advanced and morally ambiguous battles of the future will require a military composed of the highest quality individuals that can be recruited and retained. It may be that this cannot be done without some kind of inducements—positive or negative. Perhaps the military will be primarily voluntary, supplemented by some form of national service or as a requirement for college assistance.

Conclusions

Morris Janowitz emphasized the importance of a military that shares the basic values of the society it protects, even if military members cannot enjoy all the privileges of a free society while they are defending it. Whatever formal structures may be in place to ensure that the military follows the wishes of civilian society, the greatest bulwark has always been that military personnel accept the values of the liberal democratic state and have no wish to subvert them. As Justice Robert Jackson remarked in his dissent in the Japanese internment case, *Korematsu vs. United States* (323 U.S. 214):

> If the people ever let command of the war power fall into irresponsible and unscrupulous hands, the courts wield no power equal to its restraint. The chief restraint upon those who command the physical forces of the country, in the future as in the past, must be their responsibility to the political judgments of their contemporaries and to the moral judgments of history.[35]

---

34   This Vietnam era veteran can only shake his head in approving wonder.

35   *Korematsu versus United States* (323 U.S. 214), http://www.law.cornell.edu/supct/html/historics/USSC_CR_0323_0214_ZD2.html/

In a strategic environment that is becoming increasingly complex, military cohesion and effectiveness and proper civil military relations will be determined by a common bond of service and a sense of identification with a society that supports service members and values their service. This is not something new, although its importance has never been greater.

Each service has a set of core values that it expects each service member to internalize. For the Army: "duty, honor, country"; for the Navy and Marine Corps: "honor, courage, and commitment"; for the Air Force: "integrity first, service before self, and excellence in all we do"; and for the Coast Guard: "honor, respect, and devotion to duty." These military codes are fully consistent with democratic values. However the strategic environment may change, and whatever adaptations to the American way of war are required in response to these changes, they will help to ensure dedicated and honorable service in defense of the democratic state.

Chapter 5
The U.S. Civil-Military Problematique and New Military Missions

Damon Coletta[1]

When the great philosopher Plato imagined his ideal republic in the fourth century B.C.E., he saw the difficulty in arming a portion of the populace to protect his city from external threats: who then would guard the guardians? What would prevent the warrior-defenders from turning their weapons upon relatively weak and unarmed fellow citizens, overturning Plato's republic in order to take power and privilege for themselves?

Over the centuries, societies undertook dramatic experiments to address the flaws in Plato's architecture, and perhaps the greatest of these has been the shift toward a liberal-democratic constitution. In our time, this idea has swept across nearly every continent. Headlines in world affairs pivot on whether reforms in this spirit can possibly succeed for traditionally authoritarian polities in places like Russia, China, Iran, and for other important states in Africa and the Middle East. With respect to international order, the liberal-democratic Constitution served the United States exceptionally well, enabling it to grow from a loose collection of agrarian colonies huddled on the North American coast of the Atlantic to a continent-sized economic and military superpower.

For all the changes to Plato's design and despite their evident success, the answer to Plato's question of what will keep the military subservient to the state and respectful of the people it defends remains remarkably similar. Plato observed that civilization was a natural outgrowth of human nature. The central analogy underpinning his Republic likened the city to the arresting beauty and harmony of the healthy human body. The city needed intelligent leadership but not a tyrannical king, and the whole creation worked best when every piece was in its place, performing its natural function. So it was that nature bestowed the city with different types of citizens. The secret to keeping the Republic free and content was getting the right kind of people in the right positions. For the armed auxiliaries who would support the state's policy makers—the physically inferior but wiser Council

1   This is an academic work and does not represent the official opinion of the United States Government or the United States Air Force. The author would like to thank the Norwegian Institute for Defence Studies and the University of Oslo, Department of Political Science for holding seminars and providing helpful comments on previous versions of this chapter.

of Philosophers—Plato's call was surprising. The state's military officers must possess the character of noble shepherd dogs, which is to say their chief qualities were heroism and loyalty. These warriors would not trouble themselves with the higher order dilemmas of the shepherd, and they would sooner die themselves than attack the flock under their protection. Like "Man's Best Friend" through the ages, the guards would sally forth ferociously against foreign dangers yet remain steadfastly servile and gentle toward their civilian masters.

With some modifications, today's superpower Republic demands much the same from its highest ranking officers. Canine virtue, for example, while not explicitly mentioned, was implied in the standard upheld by President Obama when he explained his relief of the top commander in Afghanistan during June 2010.[2] Even so, the extraordinary durability of Plato's reply to the civil-military problematique should not be taken as a measure of the solution's efficacy but of the difficulty embedded in the question: after many trials and updates to the republican design, analysts struggle to provide a better answer. The original preoccupation motivating the civil-military problematique endures: who guards the guardians?[3]

Absent an accepted solution, national security establishments are liable to bounce around, attempting to find a way to instill that mythic sense of loyalty in their generals, consistent with military success and the country's domestic political values. For the current superpower—and a descendant of note from Plato's original Republic—the civil-military problematique is more difficult to address as economic and security challenges aggravate civil-military frictions. First, the financial and employment crises of 2008-2009 encouraged deficit spending, which, in turn, placed intense pressure on the defense budget in time of war. As resources, including personnel, equipment, and time were running low, the U.S. military dealt with a host of new missions.[4] In other words, civil-military relations are complicated now by a bumper crop of substantive issues for fueling high-stakes disputes between senior officers and their political masters.

In the modern military, these officers are not only consummate professionals but savvy agents of the Executive-branch bureaucracy. The limits on government established by a liberal-democratic constitution present irreducible opportunity for

---

2   White House, Statement by the President in the Rose Garden, June 23, 2010, www.whitehouse.gov/the-press-office/statement-president-rose-garden <March 8, 2011>.

3   Peter Feaver, "The Civil-Military Problematique: Huntington, Janowitz, and the Question of Civilian Control," *Armed Forces and Society*, 23:2 (Winter 1996): 149-78; Thomas Bruneau and Scott Tollefson (eds), *Who Guards the Guardians and How: Democratic Civil-Military Relations* (Austin, TX: University of Texas Press, 2006), pp. ix-xiii; David Pion-Berlin, "Defense Organization and Civil-Military Relations in Latin America," *Armed Forces and Society*, 35:3 (April 2009): 562-86.

4   The last chapter of Michael Horowitz's (2010) book on military innovation describes several new missions and potential investments in new capabilities under consideration by the armed services. Michael Horowitz, *The Diffusion of Military Power* (Princeton, NJ: Princeton University Press, 2010), pp. 214-25.

military leaders to slip civilian control. If the officers do not like what they hear from the President or Secretary of Defense, they can go to other decision points in a governmental system rife with checks and balances, or make an appeal through the media to abruptly drag their masters through the court of public opinion.

During the recent stress, officers have indeed availed themselves of opportunities under the Constitution to check the commander-in-chief. After the wars in Iraq and Afghanistan, practical answers to the old civil-military problematique were on the move, toward harder restrictions on civilian influence and greater military assertiveness. Several analysts have urged these changes, especially after the Bush administration's performance in the planning and early execution of the Iraq War.[5] A stronger military voice and more responsive civilian authority were corrections to keep the United States from courting national disaster in unnecessary wars.

Yet, the viability of this new military-civilian balance, like the others before, rests on Plato's incomplete answer to the problematique: noble self-restraint on the part of the warrior class who best understand and command the means of coercion. Because the right kind of abnegation, service before self, cannot be guaranteed, any civil-military arrangements based on Plato's idea are necessarily provisional. Intrusive micromanagement or draconian enforcement of civilian superiority will erode military autonomy and cripple the army's effectiveness in battle. At the other pole, unfailing deference to military claims will constrict the President until mere semantics differentiate civilian control from military management of national policy. One extreme allows no space for military virtue while the other declines to place boundaries on military mischief. Plato did not know and we cannot know precisely how much to rely on the natural nobility, the canine fidelity, of our most responsible officers. Every generation facing the problematique must feel its way; every civil-military relationship under the Republic is provisional.

That said, the pendulum in the U.S. case is swinging toward military freedom of action, perhaps due to the politics of the moment but not for scientific progress on the problematique. In anticipation of the inflection point when military prominence sets off alarm bells for a reassertion of civilian authority and bearing in mind how American civil-military relations sets a standard for a growing number of progressive republics around the globe, this critique adopts a demanding standard for military restraint. Military service takes place in a context of liberal-democratic values that tilt the playing field toward civilian authority, to the extent of accepting on principle the civilian's "right to be wrong."[6] Moreover, refocusing attention

---

5 Michael Desch, "Bush and the Generals," *Foreign Affairs*, 86:3 (May/June 2007): 97-108; Christopher Gibson, *Securing the State: Reforming the National Security Decisionmaking Process at the Civil-Military Nexus* (Aldershot: Ashgate, 2008); Dale Herspring, *Rumsfeld's Wars: The Arrogance of Power* (Lawrence, KS: University Press of Kansas, 2008); Andrew Milburn, "Breaking Ranks: Dissent and the Military Professional," *Joint Force Quarterly*, 59 (4th Quarter, 2010): 101-107.

6 Peter Feaver, *Armed Servants: Agency, Oversight, and Civil-Military Relations* (Cambridge, MA: Harvard University Press, 2003), pp. 6, 65, 265, 300.

on the relevant problematique from a political science perspective and noting the unlikelihood of resolving the civil-military relationship at once and for all time has the added benefit of pulling observers outside the rough currents of today's political rivalries.

The Problematique and American Power

The modern civil-military problematique, at least for American political science, emerged out of work by a young professor in the Eisenhower era who would go on to become one of the most famous academics of the twentieth century.[7] Samuel Huntington's *The Soldier and the State* (1957) hit on a problem that preoccupied Eisenhower himself: how the exigencies of championing the Western cause during the Cold War might slowly strangle liberty at home.[8] One of those demands, of course, was the development and provisioning of an enormous standing military. By the late-1950s, the American armed forces maintained weapons systems to win superiority in every combat environment—land, sea, and air—and in every potential theater. Missile and bomber bases sprung up to ring the Soviet Union as part of the containment strategy, and still it was not enough.[9] By the time *The Soldier and the State* came out, the Eisenhower administration's New Look, a nuclear-age force posture to stem the budgetary tide, was cracking under heavy, partisan-fueled criticism. The next President would spend even more taxpayer dollars to create a suite of options, so the U.S. military response could be supple enough to deal appropriately with whatever sort of Soviet provocation.

Would so much of the national product and national psyche devoted to military strength undermine American democracy?[10] Plato's original conundrum was coming back in spades: who or what would protect free society from its new protectors emerging from within the American Republic? The professional warrior caste would wield unprecedented raw, physical power and carry enormous bureaucratic weight in the administration of government. Huntington's questions reverberate today, though the Cold War challenge has long faded, because after

---

7   Robert Kaplan, "Looking the World in the Eye," *Atlantic Monthly* (December 2001): 68-82. Tamar Lewin, "Samuel P. Huntington, 81, is Dead," *New York Times* (December 29, 2008): B8.

8   Samuel Huntington, *The Soldier and the State: The Theory and Practice of Civil-Military Relations* (Cambridge, MA: Harvard University Press, 1957).

9   John Lewis Gaddis, *Strategies of Containment* (New York: Oxford University Press, 1982); Robert Bowie and Richard Immerman, *Waging Peace: How Eisenhower Shaped an Enduring Cold War Strategy* (New York: Oxford University Press, 1998).

10   Harold Lasswell "The Garrison State," *American Journal of Sociology*, 46:4 (January 1941): 455-68; Aaron Friedberg, *In the Shadow of the Garrison State: America's Anti-Statism and Its Cold War Grand Strategy* (Princeton, NJ: Princeton University Press, 2000); Andrew Bacevich, *Washington Rules: America's Path to Permanent War* (New York: Metropolitan Books, 2010).

9/11, American forces re-entered the fray abroad, sending ripples back to the fragile balance of power separating the branches and sustaining responsive, constitutional government in Washington.[11]

By an inspired invocation of political theory, then, Huntington would address an impressive sweep of history—across time and space—with relatively simple but powerful concepts. The usual hyperbole—magisterial, seminal, and the rest— genuinely applied to this dissertation. It would draw important critiques from sociologists and historians, but for students of government, intent upon studying institutions and culture to figure out "who decides" and with what consequences for the state, surpassing *The Soldier and the State* was like moving a mountain.[12] Huntington's book articulated the civil-military problematique for the United States as a world power, and to this day, political scientists are hard-pressed to offer an alternative.[13]

Huntington built his edifice on three pillars: the liberal values of American society, the checks and balances of American institutions, and the professionalism of American military officers. He intended his analysis of civil-military relations to be scientific, but he also wanted his work to be useful for deciding policy. Those twin desires landed him in the paradox confronted by positivists, realists, or rationalists who want to understand the world in order to make it better: if important outcomes for human beings are caused by systematic social forces, where is the space for change? Where is the room for freedom of action and moral responsibility?

Huntington's fix for the gaping hole between theory and practice is open to interpretation and part of the reason *The Soldier and the State* had such progeny. It is fair to say, though, that Huntington's pillars were true pillars because none of them changed easily. Culture, constitution, and profession were nonetheless

---

11   Norman Ornstein and Thomas Mann, "When Congress Checks Out," *Foreign Affairs*, 85:6 (November/December 2006): 67-82; Charles Kupchan, "Grand Strategy for a Divided America," *Foreign Affairs*, 86:4 (July/August 2007): 71-84; Bacevich, *Washington Rules*.

12   For a review of political science literature in the field and its connection to lines of research in other disciplines such as military sociology, see Peter Feaver, "Civil-Military Relations" *Annual Review of Political Science*, 2 (1999): 211-41, especially 211-13.

13   Several have tried, among them students who were able to study under Huntington at Harvard. I argue they did not go so far as to redefine the problematique. Rather, their recommendations for civilian control landed at different points on the spectrum of civil-military solutions established by Huntington. Michael Desch, *Civilian Control of the Military: The Changing Security Environment* (Baltimore, MD: Johns Hopkins University Press, 1999); Eliot Cohen, *Supreme Command: Soldiers, Statesmen, and Leadership in Wartime* (New York: Free Press, 2002); Bruneau and Tollefson, *Who Guards the Guardians*; Gibson, *Securing the State*; Risa Brooks, *Shaping Strategy: The Civil-Military Politics of Strategic Assessment* (Princeton, NJ: Princeton University Press, 2008); Suzanne Nielsen and Don Snider (eds), *American Civil-Military Relations: The Soldier and the State in a New Era* (Baltimore, MD: Johns Hopkins University Press, 2009).

social constructions; all of them could change over long periods of time, or with much smaller probability all at once. At various places in his volume, Huntington described what sort of changes in the three pillars—and the strategic environment—would ease the tensions between military effectiveness and civilian control, but none of the shifts were likely, none of the adjustments particularly easy to perform. Consequently, the tensions and the problematique should be with us for some time, even as civil-military stresses and ensuing policy recommendations rise and fall.[14]

Huntington discussed possible variations in military professionalism, and in a controversial epilog, the benefits to civil-military relations if individualistic, avaricious American society were to embrace some of the values on discipline and stoic sacrifice modeled at West Point. His best illustration, though, of how contending social forces pulled the actual practice of civil-military relations back and forth came with his description of institutional arrangements at the nexus of national security decision-making under the Constitution.[15]

These passages are not the most cited, even though they provided a blue-print for how a change in the external threat and subsequent interest in new military missions on the part of civilians would affect the civil-military problematique. Huntington created clear categories for different civil-military regimes, but they were challenging to operationalize. For example, it was not clear from Huntington's prose when he was sketching organizational charts—formal reporting lines and functional assignments codified by statute—and when he was discussing less

---

14   Recent civil-military relations literature reflected an awareness of the ebb and flow, originating from the socially constituted nature of Huntington's pillars. The call has been for greater mutual awareness and attention to operating style—on the part of civilian officials and soldiers—in order to keep the inevitable tensions and fluctuations in policy influence from breaking up the national security team. Dale Herspring, *The Pentagon and the Presidency: Civil-Military Relations from FDR to George W. Bush* (Lawrence, KS: University Press of Kansas, 2008); Marybeth Ulrich and Martin Cook, "U.S. Civil-Military Relations since 9/11: Issues in Ethics and Policy Development," *Journal of Military Ethics*, 5:3 (November 2006): 161-82; Desch, "Bush and the Generals;" Nielsen and Snider, *American Civil-Military Relations*; Sarah Sewall and John White, *Parameters of Partnership: U.S. Civil-Military Relations in the 21st Century* (Cambridge, MA: Harvard Kennedy School Project on Civil-Military Relations, 2009). For a treatment of new norms and practices in the broader context of civil-military relations across society as well as in the government, see Mackubin Thomas Owens, *U.S. Civil-Military Relations after 9/11: Renegotiating the Civil-Military Bargain* (New York: Continuum International, 2011).

15   Huntington, *The Soldier and the State*, pp. 186-9. The evocative phrase, "nexus of national security decision making," was paraphrased from Christopher Gibson's, *Securing the State*. It aptly identifies the central concern of the civil-military problematique and explains why the U.S. case has been treated more often by International Relations specialists rather than scholars in American or Comparative Politics. A version of the expression reappeared in Matthew Moten's *Foreign Affairs* article, "Out of Order," *Foreign Affairs*, 89:5 (September/October 2010): 2-8. See also William T.R. Fox, "Civilians, Soldiers, and American Military Policy," *World Politics*, 7:3 (April 1955): 402-18.

explicit behavior patterns, what might be called organizational culture, within the Executive. In any case, if we allow that "vertical," "coordinate," and "balanced" may refer to informal institutions, Huntington explained well why civil-military problems under constitutional democracy endure and how influence in the relationship sloshes about in response to political pressures.

Oscillating Institutions under a Constitutional Democracy

Huntington described vertical institutions as those in which the President's cabinet Secretary at the Department of Defense acted as a sort of deputy commander-in-chief. The civilian authority's scope across defense issues knew no bounds, and in practice, the civilian's superior level in the decision-making hierarchy was regularly enforced. Huntington's abstract rendering of vertical civil-military relations in 1957 uncannily captured George W. Bush's structuring of the Defense Department decades later under Donald Rumsfeld as the new President replaced the Clinton team in 2001.

Huntington also made a good prediction about what happened next, once vertical relations were established. The professional military chafed. Unlimited scope for the civilian Secretary meant senior officers' expertise on military matters came under frequent, intrusive, and unwelcome scrutiny. Military preferences were too often countermanded because regardless of the administrative or operational nature of the question, the answer tended to be mauled by the deputy commander-in-chief before it got to the President.[16]

Moreover, the American Constitution secured democratic control of defense policy and the military arm by limiting the executive commander-in-chief through checks and balances. Today, these are sometimes characterized as separate powers combined with overlapping competencies, or shared responsibilities, but this terminology does not quite capture the adversarial system promoted by James Madison in essays leading up to his culminating argument in *Federalist 51*.[17] The Founder had researched various ways, including a council of wise men, to try and prevent one governmental power, either the executive or the legislature,

---

16   After Rumsfeld and Iraq, this was the chief complaint and the inspiration for the Madisonian Approach developed by Army Colonel Christopher Gibson in *Securing the State* to give professional military voices greater weight in Presidential defense councils.

17   Geoffrey Corn and Eric Jensen, "The Political Balance of Power over the Military: Rethinking the Relationship between the Armed Forces, the President, and Congress," *Houston Law Review*, 44:3 (September 2007): 533-608 elaborated on the "shared responsibility" formulation, especially p. 557 and Section II. In recent civil-military relations literature, Feaver and Sharp used "separate but overlapping," and Moten called it "divided control over military affairs." Peter Feaver and Kristin Thompson Sharp, "The United States," in *Governing the Bomb: Civilian Controls and Democratic Accountability of Nuclear Weapons*, edited by Hans Born, Bates Gill, and Heiner Hanggi (Oxford: Oxford University Press, 2010), pp. 25-50; Moten, "Out of Order."

from encroaching upon the other. His answer in *Federalist 51* gave each branch the means to defend itself through calibrated levers of interdependence that each side could pull harder and harder as its prerogatives became endangered. "Shared responsibilities" makes it sound as if the Founders envisioned happy cooperation that might be forged in gentle deliberation between political leaders with disinterested ideas about the national good. They might have hoped for such enlightened statesmanship, but they feared Machiavellian ploys to aggrandize power. In his institutional analysis, Huntington emphasized the darker aspect of Constitutional checks-and-balances: the American President did not share his commander-in-chief competencies or his authority over the military, but the Constitutional powers of Congress to declare war, approve federal spending, and regulate the armed forces, for example, weakened the President in his corner of the civil-military relationship.

If the military should chafe, then, under vertical arrangements for Executive control, the generals could approach Congress and persuade legislators to begin tugging on the fine strings of interdependence. The military officers might expect to gain traction as well if the nub of the problem was an over-concentration of power in the hands of a deputy commander-in-chief, or cabinet Secretary of Defense. Congress, in turn, has the power to set the rules and appropriate funds for a reorganization of the Executive's Department of Defense that would formally cut away the power of the Secretary, altering reporting requirements or the span of his decision authority as it did in famous legislative watersheds of 1947 and 1986.

More likely, in practice, the credible threat of revamping departments under the President would persuade the commander-in-chief to let the steam out of military grievances by informally shifting toward what Huntington termed a coordinate system.[18] The central idea was to ensure that senior officers gain a proper hearing by removing the barrier on their level of influence. At the top of the pyramid, civil-military deliberations occurred within a V-formation, with the civilian Secretary and the military chairman occupying de facto coordinate positions on a single tier directly beneath the President.

The coordinate system delivered military professional advice directly to the President's ear, and in principle it established clear jurisdictions for the scope of civilian management under the Secretary of Defense and military autonomy under the uniformed chiefs of staff. Nevertheless, just as inferiority in the level of authority under vertical arrangements led the military to challenge the civilian deputy's competence on operational questions involving a high quotient of professional expertise, limits on the scope of authority under coordinate arrangements prompted the military to chafe once again.

This time the President, preoccupied with the wide-ranging demands of a chief executive as he simultaneously weighed military advice, struggled to maintain already fluid boundaries separating administrative from operational or strategic from tactical concerns. Huntington's rendering of the civil-military problematique

---

18   Huntington, *The Soldier and the State*, pp. 186-9.

took heavy criticism over the years for overstating the conceptual distinction between military and political decisions, but in identifying the weakness of coordinate civil-military relations, Huntington recognized that the boundary was to an important degree a construction of the players involved. He feared the military could leverage direct access to the commander-in-chief in order to strengthen its autonomy and extend the scope of its influence, calling into question the nature and significance of civilian control.

Because of the American Constitution's animating concern against the concentration of power in Congress or the chief executive, who could, if he gained the upper hand, become a kind of monarch, the military always had an avenue to push back against civilian authority. Splitting the civilians or jeopardizing the President's approval rating by going public had their risks. The military could lose its leading officers or its institutional autonomy once Congress and the President stood shoulder-to-shoulder against it, but at times under vertical or coordinate arrangements, an American military that took great pride in its professional competence and resented unnecessary civilian interference might feel it had more to gain by kicking over the table and renegotiating the norms of civilian control.[19]

The strain on the military to play for change should increase as civil-military relations move closer to the vertical or coordinate archetype. The military could use a broad array of administrative and operational issues to spread the defenses of a deputy commander-in-chief and catch him where he was weakest, undermining a distinct civilian level of authority that blocked top officers from the President; alternatively near the coordinate pole, talented military advisers or commanders could use direct access to the President in order to crowd the scope of civilian influence, pushing the boundary further away from operational issues and into areas of civilian administration or grand strategy. Huntington predicted these conditions would lead to oscillation between suboptimal civil-military systems during America's ascendance as a world power, when the professional military would remain a central feature of the national political landscape.

A more stable solution would combine best elements of the vertical and coordinate systems, limiting both the military's level and scope of authority. The President would remain in charge overall, with a Secretary of Defense to focus on political-military issues and buffer direct military pressure, but both the commander-in-chief and his cabinet Secretary would practice fastidious self-restraint, deferring to senior officers on questions of professional expertise. Huntington had little confidence that American civil-military relations could discover and settle on this sweet spot, which he called a balanced system. More likely, the fluctuating American arrangements would blast right through it on their way to either the vertical or coordinate extreme. The President was by Constitutional design simply too weak before other political actors, too vulnerable

---

19 Desch, "Bush and the Generals;" Paul Yingling, "A Failure of Generalship," *Armed Forces Journal* (May 2007), www.armedforcesjournal.com/2007/05/2635198/; Gibson, *Securing the State*.

in his ill-defined commander-in-chief role to absent himself from military councils at the same time that he granted uniformed leaders autonomy on a loosely bounded set of military issues, which promised to expand as the international threat or a national security crisis deepened.

New Missions, Old Problems

For the United States of America operating in the crucible of great power competition, Huntington articulated the civil-military problematique along three dimensions. The demands of a lead role on the world stage, particularly as post-war relations with the Soviet Union deteriorated, would keep the military near the center of American national life, commanding sufficient resources to accumulate potential for disruption in an essentially liberal political order. The likelihood and magnitude of disruption for the Republic would depend on (1) the quality of professionalism in the military, (2) the tolerance of liberal society for conservative military culture in its midst, and (3) institutional arrangements. With respect to this civil-military problematique, scholarly attention after the Cold War attended the changing nature of the external threat and primarily the first two of Huntington's dimensions.

As the Cold War threat faded, the source of new dangers became less certain and grand strategy became more difficult to articulate. Some analysts noted that since the United States still sought a leadership role in world affairs, even without Soviet armored divisions in Eastern Europe, the American military was unlikely to withdraw from its starring role in national debates. Moreover, uncertainty among civilian authorities created a natural opening for senior officers to assert more influence in national security decision-making.[20]

During the Clinton administration, the logic of a diminishing external threat seemed to play out. The relatively inexperienced President had campaigned on economic policy, but he soon confronted a series of difficult national security challenges in the Balkans, Somalia, and Haiti. Violence in those places did not imperil U.S. territorial integrity but neither could the last remaining superpower and champion of international order ignore the human suffering or the gross defiling of liberal-democratic principles in a globalizing world.

President Clinton was more or less compelled to engage the military in new peacekeeping missions. While casualties were light in comparison to the historical standard for major wars, the operations tempo during a time of defense budget draw-downs was not. The military did not always accept the new tasks passively, with their burgeoning political constraints and unfamiliar operational risks. As late as 1999, a now battle-tested administration raised the stakes, ordering a 78-day

---

20 Deborah Avant, "Are the Reluctant Warriors Out of Control: Why the U.S. Military is Averse to Responding to Post-Cold War Low-Level Threats," *Security Studies*, 6:2 (Winter 1996/1997): 51-90.

air campaign led by the supreme NATO and U.S. European commander to punish Serbia and halt ethnic cleansing in Kosovo. The post-Cold War threat environment offered plenty of opportunities for Cold War-scale civil-military conflict.[21]

The ideological dimension of the problematique received its fair share of attention during the 1990s. Though candidate Clinton had moved the Democratic Party to the center in order to capture the White House, in many ways he embodied the commercial energy and consumer-oriented brashness of Highland Falls in Huntington's famous epilog, sharply contrasting against traditional military traits of stoicism, discipline and self-sacrifice. In a youthful letter to explain his backing out of reserve officer training and perhaps a tour in Vietnam, Clinton had even paraphrased the intolerance Huntington imputed to liberal society, writing as a recent Georgetown graduate and Rhodes Scholar in 1969 that "many fine people" were "still loving their country but loathing the military."[22]

Of course, much had changed since Vietnam, the military after the 1991 Gulf War enjoyed the confidence of nearly nine out of 10 poll respondents, leading all government institutions and the press in national surveys.[23] The Reagan Revolution consolidated powerful conservative voting blocks and forced a rightward shift from the Democrats. Another conservative moment would occur early in Clinton's presidency with the rise of Newt Gingrich and House Republicans. Subsequent opinion research using data and statistical techniques unavailable to Huntington in the mid-1950s showed that individual responses were often difficult to pigeonhole into broad ideological constructs. From a social science perspective, there was scarce evidence that either liberal society or the military ethos created undue friction during the Clinton years. Even when freed from the Cold War threat, attitudes mixed across civil-military boundaries, and in any case they managed to retain at least a working level of tolerance for one another's world view.[24]

Nevertheless, a related fear highlighted ideological divisions *within* what Huntington had categorized as liberal American society. American political parties after the Nixon-era Watergate scandal and the electoral successes of the Republican

21   Richard Kohn, "Out of Control: The Crisis in Civil-Military Relations," *The National Interest*, 35 (Spring 1994): 3-17; Don Snider and Miranda Carlton-Carew (eds), *U.S. Civil-Military Relations: In Crisis or Transition?* (Washington, D.C.: Center for Strategic and International Studies, 1995); Deborah Avant, "Conflicting Indicators of 'Crisis' in American Civil-Military Relations," *Armed Forces and Society*, 24:3 (Spring 1998): 375-87; Peter Feaver, "Crisis as Shirking: An Agency Theory Explanation of the Souring of American Civil-Military Relations," *Armed Forces and Society*, 24:3 (Spring 1998): 407-34.

22   A purported copy of the letter is available at www.leatherneck.com/forums/showthread.php?t=14265 <March 10, 2011>.

23   David King and Zachary Karabell, *The Generation of Trust: Public Confidence in the U.S. Military Since Vietnam* (Washington, D.C.: American Enterprise Institute Press, 2003), pp. 4-7.

24   Darrell Driver, "Ideology and the Military Profession: A Reassessment of the Military Mind," in Nielsen and Snider, *American Civil-Military Relations*, pp. 172-93.

Southern strategy became more ideologically homogeneous. Now in the 1990s, party identification among military officers was changing as well, from neutral to Republican.[25] If a single party affiliation dominated throughout the officer corps, party loyalty could conceivably dilute military judgment when service chiefs or the field commander advised a Democratic commander-in-chief. This concern grew as professional relationships between Clinton staffers and high-ranking officers deteriorated. At one point an Air Force general, Harold Campbell, earned a rebuke and early retirement for *ad hominem* remarks against the President from the lectern at a formal banquet in Europe.[26] Notably, the accusations of gay-loving, womanizing, pot-smoking, and draft-dodging did not aim directly at Clinton's policies but at the Commander-in-Chief's moral qualifications for putting military lives in harm's way.

Overtly disrespectful behavior on the part of civilian staffers as well as prominent officers waned with time. Both sides recognized the value of getting along, and with established rotation cycles through senior military positions, more and more officers advanced through the ranks under Clinton appointees. Also, military ardor for the opposition's alternative ideas on national security would fade once Republican George W. Bush entered the White House and began to assert his understanding of civilian authority.

Nevertheless, legitimate concern over politicization of the military through lopsided party affiliation—one survey put the figure at 8-to-1 in favor of Republican officers—generated a stream of research on the importance of military professionalization during a period of ill-defined threats and ambiguous national security doctrine. Certainly, the civilians still required frank military advice, but it ought to be delivered in ways that did not prematurely constrain the President's options or trump his preferences. Civilians under democratic control had a "right to be wrong," and from a principal-agent perspective, it was problematic if the chief executive was prepared to defend a preference, say, for certain peace enforcement missions before Congress and the American people, but the military insisted otherwise.[27] As operational demands, at least for U.S. forces, wound down

---

25   Andrew Bacevich and Richard Kohn, "Grand Army of the Republicans: Has the U.S. Military Become a Partisan Force?" *The New Republic*, 217 (1997): 22-5; Ole Holsti, "A Widening Gap between the U.S. Military and Civilian Society? Some Evidence," *International Security*, 23:3 (Winter 1998/1999): 5-42; Peter Feaver and Richard Kohn, "The Gap: Soldiers, Civilians, and Their Mutual Misunderstanding," *The National Interest*, 61 (Fall 2000): 29-37.

26   John Lancaster, "General Who Mocked Clinton Set to Retire; Punishment Follows Remarks at Banquet," *Washington Post* (June 19, 1993).

27   In an otherwise well-argued and comprehensive essay on civil-military relations after Iraq, Frank Hoffman, who might have been writing from a military perspective, disagreed: "We all realize that civilian leaders have a *right to be wrong* ... but devote too little study to minimize the frequency of this occurrence" (original emphasis, pp. 232-3). This choice of words, though, may be ironic. It seems to reflect the willingness and ability of highly competent officers at the top of their profession to insist, go beyond advising, and maneuver

in the Balkans and the nation anticipated a new commander-in-chief, analysts of the civil-military problematique offered a string of appropriate behaviors that would reinvigorate military professionalism.[28]

The problem in the U.S. case, indeed for advanced democracies, was not the likelihood of a military coup but that professional officers who were also accomplished bureaucrats could make the most of institutional autonomy, and democratic limits on civilian authority, to impose military preferences on political-military questions. Professionalization merited careful attention as post-Cold War dimensions of national security rapidly changed. The concept pointed in two directions. As Huntington wrote, greater professionalization led to more autonomy, which in turn promoted greater competence among the managers of organized violence. Greater professionalization ameliorated the dilemma of civilian control, allowing both military effectiveness and democratic accountability, as long as the code included a robust ethic of self-restraint, the generals and admirals saluting smartly and executing their share of the civilian policy even when they would have decided differently. At the same time, greater professionalization meant more expertise, potentially a larger difference in perspective between military agents and civilian principals, and higher likelihood for situations when military professionals understood that if they did not insist, a less-qualified, less-informed commander-in-chief could steer the armed forces and the nation they defended off a cliff.

The dual aspect of military professionalism weakened it as a final solution to the civil-military problematique. Indeed, a parallel research thread emerged toward the end of the Clinton presidency, which turned the focus toward civil-military institutions. As in research growing out of Huntington's ideological dimension, the principal-agent approach engaged American political institutions at a different level of analysis from *The Soldier and the State*. Rather than statutory or executive reorganization of departments under the Constitution, the leading principal-agent model analyzed civilian monitoring and punishment of military shirking.[29]

The model implied that the self-restraint element of military professionalism would not prevent uniformed advisers or field commanders from straying on some

---

civilian principals onto the correct path, regardless of the principals' acknowledged authority to decide under the Constitution. Frank Hoffman, "Dereliction of Duty Redux?: Post-Iraq Civil-Military Relations," *Orbis*, 52:2 (Spring 2008): 217-35.

28   Rebecca Schiff, "Civil-Military Relations Reconsidered: A Theory of Concordance," *Armed Forces and Society*, 22:1 (Fall 1995): 7-24; Don Snider, Gayle Watkins, and Lloyd Matthews (eds), *The Future of the Army Profession* (Boston, MA: McGraw Hill, 2002); Martin Cook, "The Proper Role of Professional Military Advice in Contemporary Uses of Force," *Parameters*, 32 (Winter 2002-2003): 21-33; Suzanne Nielsen, "The Army Officer as Servant," *Military Review*, 83:1 (January/February 2003): 15-21.

29   Feaver, *Armed Servants*. Also, see commentary from several authors in a special issue of *Armed Forces and Society*, 24:3 (Spring 1998), table of contents available at http://afs.sagepub.com/content/24/3.toc <March 14, 2011>.

occasions, privileging their own policy preferences—on an issue within their area of expertise—over those of civilian authority. Empirical research into Cold War cases and civil-military frictions during the Clinton administration confirmed that pressures to shirk rose during Presidential uses of force. While the military rarely trumped civilian preferences, this commendable track record required maintenance: constant adjustments to the incentive contract for the military agent, calibrating monitoring costs, and the willingness of civilians to accept military or political risks when punishing deviation from the chief executive's program.

The principal-agent approach might have become less relevant after the Clinton years. Conservative Republican George W. Bush won the presidency in 2000 and brought with him an experienced national security team, including a Vice President, a Secretary of State, and a Secretary of Defense who should have developed a sophisticated comprehension of the code for military professionals. Rather quickly, however, civil-military relations took a different turn. Secretary of Defense Donald Rumsfeld and Secretary of State Colin Powell failed to develop a close working relationship. Rumsfeld and Vice-President Dick Cheney charged the previous administration with abusing the military in peacekeeping missions while cutting its resources. Help was on the way, Bush had promised during the campaign, but it arrived in the form of a wrenching transformation, prodding the services to adopt information technologies, so joint task forces could do more, in less time, using less mass.[30] Cutting mass fell hardest on the Army, which had parochial and legitimate reasons to resist transformation ideology. Accordingly, civilian guidance on new missions, not so much in peacekeeping as conventional warfare, pulled the military away from the successful heavy formations of Gulf War I. The elements of a rocky principal-agent relationship—incessant monitoring, micromanagement, and punishment—rushed back to the fore.[31]

The second surprise came with the September 11 terrorist attacks on the United States. The country responded comprehensively: several executive departments reformed operations, and Congress approved major reorganizations to create the Department of Homeland Security as well as a new Intelligence Community. Yet, the military was seen as one of the most capable organizations in being at the time of the attacks. During the ensuing months, the President with the approval of Congress asked it to do much more. The new missions in Afghanistan, Iraq, and elsewhere seemed to fit Rumsfeld's earlier warnings. For a brief time—through the most intense period of crisis vulnerability for the United States and the destruction of Saddam Hussein's regime in Iraq—the Secretary enjoyed clear ascendance over dissenting subordinates in the military.

---

30   Desch, "Bush and the Generals;" Robert Kaplan, "What Rumsfeld Got Right," *Atlantic* (July/August 2008), available at www.theatlantic.com/magazine/archive/2008/07/what-rumsfeld-got-right/6870/

31   Desch, "Bush and the Generals," Matthew Moten, "A Broken Dialog: Rumsfeld, Shinseki, and Civil-Military Tension," in Nielsen and Snider, *American Civil-Military Relations*, pp. 42-71.

Circumstances abruptly changed, again, as violence in Iraq swelled in the aftermath, or post-combat stage, of Operation Iraqi Freedom. Transformation targeted the source of power in adversary states, the military organizations that made dictators like Saddam untouchable and magnified their capacity for mischief during the post-Cold War years of Clinton and Bush's father. Rumsfeld's prodding may have helped the Army, indeed the joint force, become lighter, faster, and more lethal against opposing militaries, but it did not prepare the American military for nation-building once a tyrannical regime had been efficiently lopped off.

The Secretary took a firm hand the whole time. Suddenly, though, his cantankerous barbs and avuncular informality were not so charming. The cost and the crimes committed by all sides during Iraq's instability were laid at Rumsfeld's door. The power he brought to his position and his own goading style invited them there. The military, not in the form of any one person and not in a manner that was illegal or technically insubordinate, took its revenge. The announcements by retired officers—and in one politicized case the dated testimony of an Army chief-of-staff—were said by pundits and scholars to reflect judgments in the corps and among the soldiers, which Rumsfeld repressed through sheer callousness and disregard for military professionalism.[32]

In Iraq, the violence worsened, and after a bruising midterm election in 2006, the President accepted Rumsfeld's resignation, replacing him with Robert Gates, someone viewed as closer to Bush's father in courteous style as well as pragmatic approach. Secretary Gates' timing also turned out to be good. The President's national security team considered a substantive change in strategy during late-2006 and early 2007. With Rumsfeld gone, the face of the so-called "Iraq Surge" became that of the field commander, General David Petraeus, who expertly defended the plan before Congress. With few exceptions, Senators, backed by another set of military experts, did not give the surge much chance of success, but they hesitated to deny the General's request for more troops by cutting off funding. Within 12 months, the surge worked, at least to reduce the daily body count in Iraq. This, along with Gates' calm and openness to cooperation, lowered the temperature considerably on civil-military relations.[33]

Conclusion: Evaluating the Problematique

For most students of the civil-military problematique, a return to normalcy in the U.S. case came as a relief. With the arrival of Secretary Gates and President

---

32 Gibson, *Securing the State*; Herspring, *Rumsfeld's Wars*. For a cautionary reaction to this thesis, see Mackubin Thomas Owens, "Rumsfeld, the Generals, and the State of U.S. Civil-Military Relations, *Naval War College Review*, 59:4 (Autumn 2006): 68-80.

33 Hoffman, "Dereliction of Duty Redux," p. 229; Richard Kohn, "Coming Soon: A Crisis in Civil-Military Relations," *World Affairs*, 170:3 (Winter 2008): 69-80, especially p. 78; Moten, "Out of Order," p. 2-3.

Obama's subsequent decision to keep the longtime Republican in the cabinet, civil-military relations seemed to improve on every one of the problematique's dimensions. Uncompromising neo-conservative ideology gave way to pragmatism at the same time that the armed services became less overtly partisan and more open-minded about Democratic policy makers in charge of national security.[34] Discussions of professionalism took into account civilian as well as military codes and courtesies, which except for the McChrystal affair in June 2010, both sides more or less accepted.[35] In any case, Republicans, Democrats, and officers generally agreed on how the President responded to McChrystal's unguarded remarks against the administration. Finally, in terms of principal-agent dynamics, the late-Bush and early-Obama administrations, both benefiting from Gates' steady hand at the Department of Defense, adjusted the incentive contract: the military received respect and autonomy, and when it failed, senior officers were held to account. Where Rumsfeld had humiliated wayward generals, Gates fired them, a punishment that the corps understood as more in line with military custom.[36]

Despite the rehabilitation of civil-military relations after Iraq, continued reminders to remain vigilant on the meaning and practice of military professionalism, and the problematique's overall contribution to understanding what went wrong during the first six years of the Bush administration, Huntington's version of the research agenda proffered at the height of the Cold War still has relevance today. Moreover, a close reading of his institutional dimension highlights a contemporary problem for democratic civilian control.

The principal-agent approach, that is, the institutional track of the 1990s, illuminated how an advanced democracy like the United States could face serious dilemmas involving the quality of civilian control, even when the danger of a military coup was minimal.[37] However, the framework failed to anticipate deteriorating conditions during the Bush administration because it obscured the typology Huntington created, using vertical and coordinate systems. Under Huntington's "institutional constant" of Constitutional checks and balances, the

---

34   Jason Dempsey, *Our Army: Soldiers, Politics, and American Civil-Military Relations* (Princeton, NJ: Princeton University Press, 2009).

35   Ulrich and Cook, "U.S. Civil-Military Relations since 9/11;" Don Snider, *Dissent and Strategic Leadership of the Military Professions* (Carlisle Barracks, PA: Strategic Studies Institute of the Army War College, February 2008); Sewall and White, *Parameters of Partnership*; James Baker, "A Normative Code for the Long War," *Joint Force Quarterly*, 44 (1st Quarter 2009): 69-73; Moten, "A Broken Dialog;" James Burk, "Responsible Obedience by Military Professionals," in Nielsen and Snider, *American Civil-Military Relations*, pp. 149-71.

36   Moten, "Out of Order;" Richard Kohn, "Building Trust: Civil-Military Behaviors for Effective National Security," in Nielsen and Snider, *American Civil-Military Relations*, pp. 264-89. Much of the scholarship on evolving norms and practices of U.S. civil-military relations at the advisory level was synthesized in Mackubin Thomas Owens, *U.S. Civil-Military Relations after 9/11*.

37   Feaver, "Crisis as Shirking."

commander-in-chief's political vulnerability would prevent civil-military relations from ever settling into a balanced arrangement that limited both the military's level and scope of authority. The constant did not imply stasis but unending vacillation between suboptimal rules of the game. A reference back to Huntington's civil-military institutions leads to caution regarding today's interactions where they are most intense, on the course of the war in Afghanistan and on appropriate investments for U.S. Defense as commitments in Afghanistan and Iraq wind down. Huntington's theoretical narrative about how a vertical system collapses through to a coordinate system illuminates events from our headlines: the inchoate resistance against Rumsfeld from 2004-2006; the elevation of General Petraeus during the President's campaign to sell the "Iraq Surge" domestically; the assertiveness of General McChrystal, including his legally protected but politically potent response to press questions in London and the nature of his direct access to the President on the tarmac in Copenhagen; the press profile of General Petraeus in curtailing troop withdrawals from Afghanistan in 2011; and even the suspicions about military information operations to spin the civilian principal during Congressional visits to the field.[38]

If Huntington's institutional constant has held and U.S. civil-military relations just passed another cycle from vertical arrangements under Rumsfeld to informal processes that accommodate an extraordinarily influential commander such as Petraeus, Americans may anticipate yet another shift back to a circumstance where the President commissions an assertive deputy at the Pentagon to reduce military influence on national strategic decisions and bring commanders under tighter civilian control. Coordinate arrangements will work, as long as military leaders continue to make the right judgments on politically charged questions like troop numbers and operational plans. In an era of many new missions, though, it is all too possible for a prominent commander or adviser to win the domestic argument for the wrong policy.[39] In the event of a national security disaster, the chief executive's vulnerability under the Constitution will compel a response, as Huntington predicted, back to the vertical system for civilian control—despite its flaws.

Students of the civil-military problematique have skillfully employed its concepts to maintain scrutiny of a key intra-governmental relationship, even when the external threat did not force public attention to this issue and when democratic fundamentals were assured. Yet, the United States, and perhaps democratic

---

38   Thom Shanker, "General Is Said to Order Effort to Sway U.S. Lawmakers," *New York Times* (February 25, 2011): A10.

39   At the time of this writing, it is not clear what the right level of intervention is to save Libya, but the military brass have made their views known. In this case, the Secretary of Defense appears to agree, but other civilian advisers and the President may still have doubts. Despite the costs of Iraq and Afghanistan, difficult choices regarding new military missions are manifest. Thom Shanker, "U.S. Weighs Options on Air and Sea," *New York Times* (March 7, 2011): A8.

partners who closely attend the U.S. military example, will pay a strategic price if Huntington's institutional constant—the harrowing flight from vertical to coordinate arrangements and back again—remains so far outside the intellectual discourse on civil-military relations.[40]

---

40   David Pion-Berlin already employed what might be termed neoclassical institutions in comparative perspective; "Informal Civil-Military Relations in Latin America: Why Politicians and Soldiers choose Unofficial Venues," *Armed Forces and Society*, 36:3 (April 2010): 526-44. A parliamentary as opposed to Presidential democracy might suffer unique pathologies from a vertical, or integrated, arrangement: no logical avenue for the expert agent to chafe and repeated suppression of professional military advice. I thank Ida Fottland at the University of Oslo for this remark. See also Hew Strachan, "Making Strategy: Civil-Military Relations after Iraq," *Survival*, 48:3 (October 2006): 59-82.

Chapter 6

The War without a Strategy: Presidents, the Pentagon, and Problems in Civil-Military Relations since the 9/11 Attacks

C. Dale Walton

In the decade since the 9/11 attacks, the United States has been fighting a war in search of a strategy. Military operations have been continuously conducted against terrorist and insurgent groups, but the Washington has no clear vision of what it means to "win" in this struggle. Perhaps even more importantly, this directionless conflict has grossly distorted U.S. grand strategy. Indeed, the United States does not have a remotely satisfactory grand strategy in the proper sense, that being "[t]he direction and use made of any or all among the total assets of a security community in support of its policy goals as decided by politics" (Gray 2010: 18). American goals are simultaneously too vague and too narrow, focusing too much on the serious *yet nonetheless limited* threat presented by terrorists and using military power imprudently – and, in some respects, excessively – to counter that threat, while little attention is given to the larger global strategic context.

Some policymakers and commentators, particularly in the period immediately following 9/11 attacks, indicated that ending terrorism should be the U.S. goal in the GWOT, but that of course was so ambitious as to constitute strategic fantasy. Terrorism is a tool that actors find more or less useful depending on political and military circumstances; terrorism therefore will not end. It may indeed be possible to destroy specific groups, and it is even imaginable that, with the passage of time, *jihadism* will fade away as motivation for terrorism. The latter development, however, appears unlikely for many years and, in any case, there is little or no reason to suppose that the United States could bring such a result about. There is a very active moral and theological debate – or, more accurately, several simultaneous debates amongst different subgroups of Muslims – concerning *jihadism* and the morality of religious violence (Brachman 2009), but excessive U.S. government efforts to butt into that discussion probably would backfire, resulting in the discrediting of anti-*jihadist* Muslims as American stooges. In short, there is no plausible theory of U.S. victory that could be based on ending even *jihadist* terrorism specifically, much less terrorism in general.

Setting goals that truly are militarily and politically realistic would be the first step to constructing a coherent grand strategy of which counterterror strategy would be a component part. However, American leaders have proven deeply reluctant to do this, largely because it would imply – to use the phrase once common in British policymaking circles in reference to the Irish Troubles – that there is an "acceptable level of violence." Such a notion, unfortunately, is politically toxic in today's U.S. political debate, but without a realistic baseline it is impossible to set appropriate bench marks for victory. Yet, it should be noted that in many respects the United States in fact *has* been very successful in regard to counterterrorism in the years since 9/11. Most importantly, in that period there has not been a terrorist incident in the United States approaching the scale of the Oklahoma City bombing, much less the 9/11 attacks. Along the way, dozens of plotters – ranging in "terrorist competence" from the very dangerous to the hapless – have been arrested. Thus, one might reasonably say that *jihadist* violence is being successfully contained, at least insofar as U.S. soil is being protected effectively. However, at this point it still is impossible to have a lucid discussion of whether or not Washington is winning its counterterror war because there is a continued refusal to clearly state reasonable, solid criteria by which U.S. counterterrorist efforts will be judged.

This is important, as such skittishness makes it impossible to construct a cogent connection between ends and means. Instead, the U.S. counterterror effort is chopped into tiny individual pieces that do not form a coherent whole – on one day, rough justice is meted out to Bin Laden and the United States is "winning," but perhaps on the next a suicide bomber detonates himself in a shopping mall and the U.S. is "losing." *Without a realistic set of ends supported by appropriate means, U.S. counterterrorism cannot be a truly strategic enterprise, regardless of how brilliant it might be tactically and operationally.*

The Bush Administration does deserve at least a little credit for implicitly acknowledging this problem relatively early in the GWOT era and attempting, albeit half-heartedly, to unravel it. A short section in the *National Strategy for Combating Terrorism* entitled "Victory in the War Against Terror" (White House 2003: 12) states that:

> Victory against terrorism will not occur as a single, defining moment ... However, through the sustained effort to compress the scope and capability of terrorist organizations, isolate them regionally, and destroy them within state borders, the United States and its friends and allies will secure a world in which our children can live free from fear and where the threat of terrorist attacks does not define our daily lives. Victory, therefore, will be secured only as long as the United States and the international community maintain their vigilance and work tirelessly to prevent terrorists from inflicting horrors like those of September 11, 2001.

This declaration at least recognizes the obvious fact that all use of terrorism cannot be ended. Moreover, one might reasonably read into it a *very* oblique acknowledgement that the U.S. can be successful at counterterrorism but still suffer a limited degree of terrorist violence on its own soil. The latter interpretation, however, is made problematic by the sentimentalist language about children living free from fear and, in any event, what an acceptable level of violence would be is left extremely vague.

In the years since the *National Strategy to Combat Terrorism* was initially released, U.S. policy guidance on counterterrorism objectives has not become appreciably clearer. (Indeed, both the latter document and the periodically-updated *National Security Strategy*, are misnamed; they essentially are policy documents listing broad goals, not strategies for obtaining those goals.) Even more importantly, however, how counterterrorism fits into U.S. grand strategy has never been clarified satisfactorily. As a result, counterterrorism broadly, and the occupations of Iraq and Afghanistan in particular, have been the focus of Washington's attention for the last decade. However, as is detailed below, counterterrorism is far too narrow a focus for a superpower with global security objectives that require shaping the international system in a manner that it finds amenable.

The One-Way Street: Civil-Military Relations in the Present

U.S. deficiencies in regard to grand strategy partly are the result of a major flaw in the current U.S. civil-military relationship that is detailed below. This is not, by any means, the *only* reason why the United States has underperformed strategically in recent years, but it is an important one with obvious relevance to this volume. Moreover, it is an eminently correctable one, requiring a change in the culture of the U.S. officer corps that, while significant, is far from unimaginable. Indeed, in a relatively brief time, this shift could be cemented in U.S. military culture, *if* a critical mass of senior leaders were willing to embrace it boldly and accept the career risks that would accompany its practice.

In recent decades, U.S. military officers have come to see their duty in overly narrow terms, as mere operators who achieve whatever short-term military end is commanded by their civilian superiors. This, unquestionably, is part of their job – indeed, this is the one function that they clearly are obligated legally to perform. However, this understanding – which certainly has been encouraged by civilian policymakers, most of whom prefer a passive officer corps which does not "make trouble" – is overly narrow. Military personnel should challenge their civilian superiors intellectually, constantly probing the strategic vision of the latter, and compelling those superiors to confront the potential weaknesses of that vision.

This of course does not mean that uniformed officers should regard themselves as free to choose whether or not to implement that strategic vision – so long as they do not resign, they are obligated to implement, to the best of their ability, even a strategy that they consider unwise. However, probing civilian superiors intellectually is not insubordination. One can understand why many officers find it difficult to do so, however; all military branches emphasize the importance of immediate and unquestioned obedience to lawful orders for the quite compelling reason that combat conditions are not conducive to seminar discussion. The policy process, however, effectively *is* a never-ending seminar, albeit one in which firm decisions, good or bad, are sometimes made.

Military discomfort with a policymaking role is understandable, but there is a meaningful distinction between active participation in the policy process and actually making high-level political-strategic decisions. (As a practical matter, military officers, particularly Combatant Commanders, regularly do make political-strategic decisions of a more modest kind [Belote 2004, Priest 2003] – though neither military nor civilian decision-makers are eager to acknowledge that fact.) Certainly, it would not be appropriate for serving officers to set U.S. foreign policy priorities. However, it is legally and ethically unproblematic for military officers to speak up about their concerns and point out potential weaknesses in the strategic logic of their civilian superiors and thus (ideally) prevent them from embarking on folly or (if need be) convincing them to abandon an unwise course on which they already have embarked. Indeed, military officers have a duty to ensure that the interests of the United States are protected, and they should not watch silently as avoidable tragedies unfold. Military officers cannot veto the decisions of civilian superiors; they should do as much as is practically possible to oblige the latter to craft coherent policy goals and a grand strategy that uses military power prudently to further national objectives.

The U.S. military indisputably is subordinate to the civilian leadership in the Pentagon and, ultimately, the president. This does not, however, mean that the Constitution, and applicable federal statutes that shape the daily functioning of the civil-military relationship, requires unquestioning servility on the part of a military officer. Clearly, there is a legal obligation on the part of officers to disobey illegal orders, but the problem under discussion herein is far more subtle than a heroic decision to refuse to commit a war crime. Rather, the question is if military personnel – in this context, mostly very high-ranking general officers – have an obligation to vigorously (though not publicly) question, elbow, and confront their civilian superiors so as to encourage the latter to sharpen their own strategic thinking. It is not argued herein that three- and four-star officers necessarily should threaten to resign individually, much less *en masse* – an act that many officers find vaguely disreputable, as it has a whiff of organized mutiny, though there may be a good case for it in certain exceptional circumstances (McMaster 1997). Rather, this essay simply contends that, as a group, the military's highest-ranking officers *have proven far too reluctant to challenge their civilian superiors intellectually* and that this, in turn, has encouraged U.S. policy and grand strategy to be vague and ill-designed.

This trend has been developing at least since the end of the Second World War, perhaps the last American conflict in which military and civilian policymakers continually engaged in a very fruitful dialogue on the link between strategic ends and means, and it is likely that the post-war formation of the Defense Department has been a key factor encouraging this problem. The current DoD structure was created largely with the intention of asserting civilian supremacy in the Pentagon (Kinnard 1980), and in that respect it succeeded too well. In principle, a system with four individual service chiefs and a Chairman of the Joint Chiefs of Staff (CJCS) would appear to be an adequate one for ensuring that advice representing that the considered views of sea, air, and land warriors is readily accessible to the president and his or her staff. However, in practice it has proven possible for tough-minded secretaries of defense to utterly dominate the JCS. The result is that service chiefs are somewhat like CEOs of subordinate companies within a giant conglomerate, powerful in their own backyard but with little voice in the overall direction of the enterprise and virtually cut off from their supreme commander. The CJCS has greater access to the president, but those who hold this office sometimes have served less as independent actors advising the president than as the defense secretary's "ambassador" to the uniformed military. There are exceptions, such as Colin Powell, who, as National Security Advisor and later CJCS, blurred the distinction between civilian policymakers and general officers (Cohen 1995: 108). More usual recent occupants of the CJCS position, however, have been accommodating personalities such as Generals Richard Myers and Peter Pace. Interestingly, the former allegedly was picked by Secretary Rumsfeld for that job over Admiral Vern Clark at least partly because Rumsfeld believed Clark to be too strong-willed and frank (Graham 2009: 275).

The sidelining of the Executive Branch's military advisors of course would not be problematic if presidents and defense secretaries consistently demonstrated strategic brilliance – Clausewitzian genius trumps the formal policymaking process. It would be difficult to argue that this has been the case throughout recent decades; in the GWOT era in particular, the strategic judgment of figures including President Bush (Hoyle 2008, Record 2010) and Secretary Rumsfeld (Herspring 2008, Ricks 2006) has been much-questioned. Of course, evaluations of major political figures should not be accepted casually – certainly, some detractors of Bush and Rumsfeld have issued histrionic polemics which are so wild as to be of little or no value (Vidal 2002, Wolf 2007). However, more measured critics have made many good points about the strategic failings of these two key policymakers. Even leaving specific personalities aside, however, the U.S. process for the making of policy and strategy clearly has produced suboptimal outputs, a fact made obvious in thoughtful assessments such as that offered by Steven J. Metz (2008).

Overall, it very much appears the relationship between military and civilian policymakers has become grossly distorted. Of course, that relationship is inherently unequal, given the Constitutional principle of civilian supremacy and the statutes governing the Department of Defense. However, as alluded to above, there is nothing in the law requiring that the senior military be passive intellectually – rather than unquestioning obedience of civilian authorities, the officer corps should cultivate "inquiring obedience," particularly for officers at the most senior ranks.

Inquiring Obedience: Balancing Deference and Responsibility

Although the loyalty of the American military to civilian authority should never be in question, as a practical matter there is very little danger that the military will, at least at any point in the foreseeable future, threaten the U.S. Constitutional order. Writing in the early 1990s, one author presented a cleverly written – and, in military circles, much-discussed – dystopian future scenario in *Parameters* based on a *coup d'état* taking place in the United States in 2012 (Dunlop 2010-2011). Fortunately, however, it is likely that the year 2012 will pass quietly insofar as coups against the U.S. government are concerned.

A lesser, but far more plausible, danger is presented by a sometimes overly-close relationship between senior military officers and the media, with interviews – and at times, more problematically, leaking – used tactically to nudge policy in one direction or another. This certainly occurs regularly, but there is no way to entirely prevent this in a free society. The interaction between the press and the uniformed military ultimately is not far different today than during the Second World War, or even the American Civil War. As in the past, "policy nudging" is formally discouraged – it can even technically be criminal under some circumstances – but, in general, it is a not a critical matter; there is little threat that civilian policymakers might become the puppets of media-savvy generals. Indeed, the removal in 2010 of General Stanley McChrystal from command in Afghanistan dramatically demonstrated just how just dangerous media contacts can be for serving officers. While McChrystal and his staff essentially were accused of nothing more serious than loose and disrespectful talk, subsequent investigation has indicated that even many of the claims made in that regard were dubious (Shanker 2011). Of course, that vindication can only provide very cold comfort for those whose military careers were harmed – and in McChrystal's case, ended – by the allegations. By way of historical comparison, the machinations of a figure like Nelson A. Miles, the Commanding General of the Army at the turn of the twentieth century (Stevenson 2006: 142-5), presented a far more blatant challenge to civilian authority than anything claimed in the media reports regarding McChrystal and his staff.

Few U.S. civilian policymakers have any real academic background in the making of strategy. Strategic studies is a very small field in the United States, and only a limited number of universities offer strategic studies courses as such. Some students are exposed to strategic studies material in political science or international relations courses, but this is quite scattershot – many undergraduate students majoring in such fields never hear anything of strategic studies beyond a couple of brief references to Thucydides and Clausewitz. Yet, modern international relations theory likely is of little utility in actual policymaking. (It seems that terms such as "Constructivist" and "Neo-Realist" are used infrequently, at best, in the White House Situation Room.) In short, civilian policymakers often (but not invariably) have considerable personal experience on which to draw, but lack an intellectual background that encourages them to "think grand strategically" rather than responding to issues individually, which results in an undisciplined, ad hoc approach to strategy and policy. As Colin S. Gray notes, "It is relatively easy, albeit perilous, to fail to notice that one lacks a strategy" (2010: 130).

In contrast – rather ironically, given the military's marginalization in U.S. strategic decision-making – most senior military officers have a considerable background in strategic studies, including postgraduate education at one of the war colleges. Thus, they have a deep familiarity with key concepts and – if their strategic education has not been mere wasted time – they should know the right sorts of questions to ask of civilian policymakers who wish to use (or threaten to use) violence to obtain U.S. policy goals. At the most basic level, they should be able to look beyond immediate pressure to "do something," the result of which often is participation in dubious enterprises (a very recent example being the 2011 NATO war against Libya). It is just as important, if not more so, that they also be ready to challenge the more structured and ambitious – but perhaps catastrophically flawed – strategic notions of their civilian peers.

This is not to claim that the wisdom offered by senior military officers is a magic tonic that would solve all U.S. strategic woes. Strategy is a deeply difficult enterprise (Gray 2009; Gray 2010), and putting on a uniform with an impressive set of ribbons does not automatically make one into Fredrick the Great. However, the combination of appropriate strategic education and practical military experience can be very powerful in disciplining the strategic imagination, discouraging unhelpful flights of fancy and reminding one that what Gray calls "the strategy bridge" is not merely built out of unjustified hope.

The Intellectually Undisciplined GWOT

It is striking how rarely very basic questions appeared in the early months following the 9/11 attacks, the most obvious being what, precisely, the United States planned actually *to do* with Afghanistan once it had hunted down al Qaeda fighters and overthrown the Taliban. Ruling Afghanistan was not, after all, the only option available to the United States – and the liberal interventionist notion that outside powers have a responsibility to "fix" any country in which they act militarily can, if acted upon, easily bring about disaster (Walton 2009). The most straightforward alternative would have been simply to conduct a grand raid whose purpose would be to displace the Taliban government and capture or kill as many Taliban and al Qaeda fighters as possible, with Osama Bin Laden being (for obvious reasons) by far the most important individual target.

In this light, while the wisdom of specific military choices, such as the decision to pause combat operations during the Battle of Tora Bora during the winter of 2001-2002, may be debated, the general character of American operations during the early weeks of the Afghan conflict was entirely compatible with a raiding strategy. If the United States made a fundamental error in that period, it was not an operational one, but the political-strategic one to accept a task – the creation of a reformed Afghanistan – that would be extraordinarily difficult (if not impossible), expensive, and would guarantee that Washington could not show its friends and enemies alike that it was capable of winning a speedy and decisive victory *on its own terms* over terrorists and their allies.

The subversive thought that Washington perhaps simply could declare victory and walk away from Afghanistan, putting its client king Hamad Karzai in place and supporting him to the level required to prevent a reorganized Taliban from accomplishing any embarrassing future achievements (such as retaking Kabul, for example) appears never to have deeply penetrated the upper ranks of the Bush Administration. This, itself, is interesting, as it demonstrates how thoroughly a particular, narrow view of how the GWOT must be conducted had captured the imagination of civilian policymakers.

The notion that Afghanistan must be reordered so as to prevent its serving as an al Qaeda base was accepted with little attention to the obvious Clausewitzian question: *Was Afghanistan truly the enemy's center of gravity?* There actually was little reason to believe that it was: the world is filled with feeble and failed states, after all, and, moreover, many of these are majority-Muslim. The was no particular reason why international *jihadists* in general, much less Arab-dominated al Qaeda in particular, had to base themselves in Afghanistan. Unsurprisingly, once operating out of Afghanistan became untenable, al Qaeda members, including Osama Bin Laden himself, simply crossed over into Pakistan; in due course the organization reconstituted itself, albeit perhaps in a much looser form.

It appears that in the immediate post-9/11 environment the U.S. military failed to guide American civilian leaders to ask difficult Clausewitzian questions, an impression that is only reinforced by the memoir of the Gen. Tommy Franks (Franks 2004), at the time the commander of CENTCOM and, aside from the CJCS, the officer best situated to probe and challenge the strategic concept developing in Washington. The result was an unfocused effort and a president who concluded that the occupation of Afghanistan was a key to proving U.S. resolve in the fight against al Qaeda (Bush 2010: 191).

The decision to occupy Afghanistan, unwise though it might have been, clearly followed from the 9/11 attacks. Many have argued that the 2003 Iraq invasion, in contrast, was a separate enterprise from counterterrorist efforts aimed at al Qaeda and *jihadist* groups. After all, despite occasional obnoxious displays of pseudo-piety, such as commissioning a copy of the Qur'an written in his own blood, Saddam Hussein in essence was a secular figure. However, the Iraq War was not separate from the GWOT in the understanding of the president himself (Bush 2010: 189-90), who saw it as a major component in the (legitimately important) effort to keep weapons of mass destruction out of *jihadi* hands. Moreover, *jihadism* itself supposedly would be undermined by inserting stable democracy into Mesopotamia for the first time in human history. These notions were always problematic – the expected WMD stockpiles apparently did not exist, while, as the events following the Arab Spring continue to demonstrate, the undermining of the decrepit authoritarian Arab political order carries great dangers of its own. Nevertheless, the Iraq invasion clearly fits comfortably in the context of the overall Bush-era conduct of the GWOT.

In regard to Iraq, there was "pushback" from the senior military in the period before the invasion, but this largely was quashed by Secretary Donald Rumsfeld

(Herspring 2010: 77-90). In any case, however, uniformed resistance to the invasion focused mainly on its mechanics – how many troops would be required for the occupation and so forth. The military did not effectively press civilian policymakers to consider deeper questions of just how Iraq *realistically* would fit into their overall GWOT strategy, particularly if the occupation period were to prove far more difficult than anticipated. Most of the reasons given for the Iraq enterprise were defensible individually – key assumptions regarding WMD were quite incorrect, but that was not knowable with certainty until after the invasion was untaken; moreover, there certainly was consideration of how operations in Iraq would fit into the larger U.S. strategy to confront terrorism and related problems (Feith 2008; Mann 2004; Rumsfeld 2011; Woodward 2004).

The occupations of Iraq and Afghanistan – as well as the myriad lesser U.S. counterterror efforts worldwide – ensured that the GWOT would become Washington's obsession for years. This, in turn, blinded it to fundamental changes that were occurring in the global system.

Thinking About Grand Strategy in a Changing Global System

It is depressingly easy for leaders confronted with daily challenges of constructing U.S. foreign policy to fall into the trap of becoming fixated on the obvious problems of the moment, whatever they might be. However, this sort of thinking is inimical to good strategic thinking, as it tends to obscure larger issues. During the Cold War period, the overarching structure provided by containment proved a formidable defense against the vices that later would be demonstrated in GWOT policymaking.

A general tendency toward moderation helped to "smooth out" grand strategy, preventing one or more key strategic errors from metastasizing and coming to play a dominant, even defining, role in American strategy. The open character of U.S. society and political life surely was largely responsible for this phenomenon – criticism of U.S. policy was continual, and it came from many sources across the political spectrum. While much of this criticism was misguided, ill-informed, or even vapid, constant and public critique of government policy did encourage a healthy examination of received wisdom. Intellectual laziness certainly was not eliminated – notably, the degree to which the logic of Mutually Assured Destruction (MAD) discouraged serious thinking about nuclear conflict was astonishing – but in general it was difficult for policy makers to sustain policies based on strategic fantasy, no matter how badly they might have wished to do so. At times, unreasonable optimism did take temporary control of strategy – as in the Johnson Administration, where the belief reigned that the graduated use of force in Vietnam would provide for a quick, easy victory once the appropriate level of "pain" was inflicted on North Vietnam. In general, however, containment of the USSR was based on calculations of long-term sustainability; the United States did not need to win the Cold War by a date certain, it merely needed to endure and prevent the catastrophic erosion of its position in the world. From this perspective, U.S. policy was not particularly exciting, but it had the greater virtue of being realistic and, at least to a limited degree, self-correcting – highly valuable traits, given that its fundamental purpose was to prevent both great power war and the Soviet domination of Eurasia.

The GWOT presented U.S. policymakers with some broadly similar challenges – most notably, it was a conflict without a clear "sell-by" date and one whose ultimate outcome could not be a straightforward military victory – but it was radically different in other critical respects, most importantly in that it was not a bipolar superpower struggle whose outcome would reshape the international system. Islamist terrorist groups sometimes receive state support; al Qaeda, operating in an exceptionally feeble Afghanistan, even managed to become something of a force in the Afghan state apparatus (Coll 2004; Wright 2006), but no such group can command the resources of a superpower and there is no plausible scenario in which one might do so. Moreover, in most cultures the appeal of Islamist ideology is quite limited; in contrast, for much of the twentieth century universalistic Leninist ideology, backed by Soviet military-industrial power, was making a solid bid for global dominance (Walton 2002, Walton 2007a: 205-207).

Given these differences, it was sensible to place the containment of communism at the center of U.S. grand strategy, but later putting counterterrorism in that same position was inappropriate. A subtler, but, in the long view, far more important challenge for U.S. security was occurring during the first decade of the twenty-first century: global multipolarity was in the process of reemerging (Walton 2007b). By the time of the 9/11 attacks it was becoming increasingly obvious that China, at least, was a credible candidate for great power status, while India was moving in that direction, and Russia was quite serious about reorganizing internally so as to reverse its geopolitical freefall.

It is impossible reliably to predict the long-term future, but grand strategy – if it is practiced in any form worthy of the name – requires placing intelligent wagers on how history's arc will bend. There was every reason to believe that unipolarity could not be sustained permanently; this, in turn would mean that the international system would face new stresses and possible undesirable developments, including the possibility of great power war at some unknown future point. The United States not only did very little to address this potentially survival-level threat, but did not even pay it much intellectual attention. Rather than asking what needed to be done to shape the global environment in a manner that would serve U.S. security over coming decades, GWOT-era policymakers chiefly confided themselves to addressing the immediate danger presented by relatively weak Islamist terrorists who could not, in any event, be completely nullified as a potential threat.

The contrast with the mid-twentieth century is stark. During and after the Second World War, the United States undertook a variety of measures that shaped the security context in which the Cold War took place. Some critically important institutions – such as the United Nations and the organizations supporting the Bretton Woods system – were created even before it became clear to Washington that there would be a long-term Cold War between itself and Moscow. Other organizations – most crucially, NATO – developed later, after the security landscape was better defined. The United States certainly made strategic errors during this period; for example, the United Nations never played the role of effective guarantor of global peace that Washington envisioned it would, and, indeed, in recent decades various UN entities, notably including the General Assembly and Secretariat, often have proven hostile to U.S. foreign policy (and, arguably, to the United States itself). However, it is clear that Washington made a serious effort in the mid-twentieth century to shape the global security environment to its benefit, and that later paid substantial strategic dividends.

The result of the "GWOT focus" was the loss of 10 critical years in which U.S. power was still so great that it could have taken the lead role in shaping a new set of global institutions that might simultaneously both protect vital U.S. interests and diminish the likelihood of a future great power confrontation. However, such possibilities were not seriously explored. Instead, the United States ignored the problem of large-scale geopolitical change, content with NATO expansion and the occasional, and very half-hearted, discussion of the reform of global institutions such as the UN Security Council. The latter institution, it should be noted, probably literally is un-reformable in terms of making it a useful instrument for the management of great power relations. However, the United States government paid little mind to the question of whether a new, less formal and public but more substantive institution for the management of great power relations – a twenty-first century Congress of Vienna, so to speak – might be created.

It would have been a rather simple matter to arrive at sensible *core* U.S. strategic goals for the GWOT: first, the prevention of mass casualty terrorist assaults similar to, or greater in scale than, the 9/11 attacks against the U.S. homeland; second, the prevention of such attacks on a small number of close U.S. friends and allies, including Australia, Israel, Mexico, New Zealand, South Korea, and the NATO states. Terrorist use of WMD, particularly nuclear weapons, would be the most plausible fashion in which an attack on such a scale may successfully be undertaken. The (notably, non-WMD) 9/11 attacks most likely were a singular event made possible by weak airport security, the use of an entirely new terrorist tactic (using a passenger airliner as a weapon aimed at a ground target), and simple luck on al Qaeda's part. It admittedly is not impossible that terrorists will be able to devise an equally devastating unorthodox attack in the future, but it is quite unlikely; it is far more plausible that WMD would be the critical enabler for an attack on a massive scale. The explicit, limited objective of preventing "future 9/11s" in specific countries in turn lends itself to the crafting of a realistic strategy that is financially and politically sustainable over the long term.

Setting a specific, limited counterterrorism goal of course is not a panacea that solves all problems. Notably, the war in Iraq still could have been supported by policymakers arguing that the overthrow of Saddam Hussein was critical to preventing terrorist acquisition of chemical and biological weapons. However, cutting away excessively ambitious goals such as implanting democracy in the Middle East and thus, allegedly, undermining the conditions that encourage Islamist terrorism would have placed useful boundaries on U.S. strategic discourse – and possibly would have a nudged leaders down a different path than the one that they actually did pursue.

Conclusion: Getting Past a "Terror-centric" Grand Strategy

The death of Osama bin Laden provided an appropriate coda for a decade-long counterterrorist chapter that began on 11 September 2001 and continued until Bin Laden's encounter with SEAL Team Six on 2 May 2011. Although for most of that time Bin Laden apparently had little operational control over his al Qaeda network, the 9/11 attacks placed counterterrorism at the center of U.S. grand strategy and set in motion a series of foreign policy choices, some of which were deeply damaging to American power, consuming vast financial resources and speeding the collapse of a unipolar system in which Washington was overwhelmingly powerful. Monstrous though the 9/11 attacks were, counterterrorism always was an overly narrow focus for a superpower with global interests and which historical circumstance had burdened with the thankless task of trying to maintain some reasonable degree of peace and good order in the global system.

Now that Bin Laden is dead, the U.S. has ceased to play an active combat role in Iraq, and the withdrawal from Afghanistan is underway, it would be highly beneficial for the United States to close the book definitively on the "GWOT era" and clearly move beyond the myopic grand strategy of the past decade, instead adopting one that is focused mainly on the future of the international system and, particularly, the relationship amongst the great powers.

Unfortunately, there is little sign yet that the Obama Administration has come to have the broadness of vision needed to reorient U.S. grand strategy radically. However, it is possible that it will come to develop a truly panoramic view, but it will not do so if its top civilian policymakers are not challenged intellectually. That is a task that senior officers of the uniformed military can, and should, undertake. The grand strategy of the last 10 years has served the United States poorly. Indeed, most of the costs for ignoring the most important developments in the global system have not yet come due – there is no escaping the fact that Washington will pay some price tomorrow for mistakes made in the past. However, happily, there is little persuasive evidence that a catastrophic breakdown in the global system already has become inevitable. The objective now should be to adopt a grand strategy that will advance U.S. interests and minimize the dangers of a future calamity such as a great power war.

Bibliography

Belote, H.D. 2004. Procounsels, Pretenders, or Professionals? The Political Role of Regional Combatant Commanders, in Office of the Chairman of the Joint Chiefs of Staff, *Essays 2004: Chairman of the Joint Chiefs of Staff Strategy Essay Competition.* Washington, DC: National Defense University Press: 1-20. Available at http://www.isn.ethz.ch/isn/Digital-Library/Publications/Detail/?ots591=0c54e3b3-1e9c-be1e-2c24-a6a8c7060233&lng=en&id=100788 [accessed: 24 May 2011].

Brachman, J.M. 2009. *Global Jihadism: Theory and Practice.* London: Routledge.

Bush, G.W. 2010. *Decision Points.* New York: Random House.

Cohen, E.A. 1995. Playing Powell Politics: The General's Zest for Power. Review essay. *Foreign Affairs,* 74:6: 102-10.

Coll, S. 2004. *Ghost Wars: The Secret History of the CIA, Afghanistan, and Bin Laden, from the Soviet Invasion to September 10, 2001.* New York: Penguin Press.

Dunlop, Jr., C.J. The Origins of the American Military Coup of 2012. *Parameters* 40(4): 2-20. Originally published in *Parameters,* 23:4.

Feith, D.J. *War and Decision: Inside the Pentagon at the Dawn of the War on Terrorism.* New York: Harper.

Graham, B. 2009. *By His Own Rules: The Ambitions, Successes, and Ultimate Failures of Donald Rumsfeld.* New York: PublicAffairs.

Gray, C.S. 2009. *Schools for Strategy: Teaching Strategy for the 21st Century.* Carlisle, PA: Strategic Studies Institute/U.S. Army War College.

Gray, C.S. 2010. *The Strategy Bridge: Theory For Practice.* New York: Oxford University Press.

Franks, T. 2004. *American Soldier.* New York: HarperCollins.

Herspring, D.R. 2008. *Rumsfeld's Wars: The Arrogance of Power.* Lawrence, KS: University Press of Kansas.

Herspring, D.R. 2010. Rumsfeld as Secretary of Defense, in *The George W. Bush Defense Program: Policy, Strategy, & War,* edited by S.J. Cimbala. Washington, D.C.: Potomac Books.

Hoyle, R. 2008. *Going to War: How Misinformation, Disinformation, and Arrogance Led America Into Iraq.* New York: Thomas Dunne.

Kinnard, D. 1980. *The Secretary of Defense.* Lexington, KY: University Press of Kentucky.

Mann, J. 2004. *Rise of the Vulcans: The History of Bush's War Cabinet.* New York: Viking.

McMaster, H.R. 1997. *Dereliction of Duty: Lyndon Johnson, Robert McNamara, the Joint Chiefs of Staff, and the Lies that Led to Vietnam.* New York: HarperCollins.

Metz, S. 2009. *Iraq and the Evolution of American Strategy.* Washington, DC: Potomac Books.

Stevenson, C.A. 2006. *Warriors and Politicians: US Civil-Military Relations Under Stress*. London: Routledge.

Priest, D. 2003. *The Mission: Waging War and Keeping Peace with America's Military*. New York: W.W. Norton.

Record, J. 2010. *Wanting War: Why the Bush Administration Invaded Iraq*. Washington, D.C.: Potomac Books.

Ricks, T.E. 2006. *Fiasco: The American Military Adventure in Iraq*. New York: Penguin Press.

Ricks, T.E. 2009. *The Gamble: General David Petraeus and the American Military Adventure in Iraq, 2006-2008*. New York: Penguin Press.

Rumsfeld, D. *Known and Unknown: A Memoir*. New York: Sentinel.

Shanker, T. 2011. Pentagon Inquiry into Article Clears McChrystal and Aides. *New York Times* [online 18 April]. Available at http://www.nytimes.com/2011/04/19/us/politics/19military.html?scp=3&sq=mcchrystal percent20AND percent20rolling percent20stone&st=cse [accessed: 20 May 2011].

Vidal, G. 2002. *Dreaming War: Blood for Oil and the Bush-Cheney Junta*. New York: Nation Books.

Walton, C.D. 2002. Catastrophic Errors: Totalitarian Ideology in the Twentieth Century. Review essay. *Comparative Strategy*, 21:2: 115-19.

Walton, C.D. 2007a. Triumph Over Illusion: William R. Van Cleave and the Battle Against Conventional Wisdom, in *American National Security Policy: Essays in Honor of William R. Van Cleave*, edited by B.A. Thayer. Fairfax, VA: National Institute Press, 205-209.

Walton, C.D. 2007b. *Geopolitics and the Great Powers in the Twenty-First Century: Multipolarity and the Revolution in Strategic Perspective*. London: Routledge.

Walton, C.D. 2009. The Case for Strategic Traditionalism: War, National Interest and Liberal Peacebuilding. *International Peacekeeping*, 16:5: 717-34.

White House. 2003. *National Strategy for Combating Terrorism*. Available at https://www.cia.gov/news-information/cia-the-war-on-terrorism/Counter_Terrorism_Strategy.pdf [accessed 10 April 2011].

Wolf, N. 2007. *The End of America: Letter of Warning to a Young Patriot*. White River Junction, VT: Chelsea Green.

Woodward, B. 2004. *Plan of Attack*. New York: Simon & Schuster.

Wright, L. 2006. *The Looming Tower: Al-Qaeda and the Road to 9/11*. New York: Knopf.

Chapter 7

Who Serves? The American All-Volunteer Force

Gary Schaub, Jr. and Adam Lowther

Who Serves?

Who serves in the military? When the United States ended conscription and began to acquire its military personnel voluntarily, significant concerns were voiced. Would the military attract sufficient and appropriate personnel? Would the self-selected force reflect American society in terms of demographics, socio-economic origin, and ideology? Or would the force become increasingly separate and alienated from American society, maneuver to become politically independent from civil authority, and perhaps endanger the polity? We address these issues by discussing the underlying choice made by the U.S. government when it opted for an all-volunteer force, reviewing many of the concerns raised about consequences of this choice, assessing the degree to which these occurred and whether they still affect the force, and discussing concerns raised about the current force by the leadership of the Department of Defense.

Equity of Process versus Equity of Outcome

Who should decide who bears the burden of military service: the government or the individual? Two diametrically opposed answers are compatible with different strains of democratic theory: civic republicanism and classical liberalism. The former emphasizes the benefits of coerced civic participation and downplays its inefficiencies while the latter stresses the efficiency of the market and de-emphasizes its impact on civic life.

In civic republican thought, emphasis is placed upon developing individuals as citizens and engaging them in public life—including governance and defense. As such, it is right and proper to use the coercive power of the state to ensure *universal participation* of citizens in the military. Participation has many salient effects. First, it enlarges the interests of individuals to include those of the community. It encourages them to value the common good of the polity and provides them with a sense of ownership and responsibility for its well-being. Wide-spread participation in civic life is presumed to guard the polity against the tyranny of narrow interests. It also assists harmonious civil-military relations as induction into the military provides the state the opportunity to indoctrinate the populace in patriotism as well as martial virtues (which, hopefully, are positive). It therefore encourages a convergence in the preferences of the state and society, thereby reducing the chances that the military will use its power against the state.[1]

Yet, from the perspective of democratic theory, compulsory service to the state has many problems. First, it constitutes a deprivation of liberty—involuntary servitude to the state.[2] As Milton Freidman put it, "One of the great gains in the progress of civilization was the elimination of the power of the noble or the sovereign to exact compulsory servitude."[3] Such deprivation could be seen as obligatory service to the community if it applied universally to all. Yet, historically, it has not.[4] It also has not been necessary as it has rarely been the case that modern states have required universal military service.[5] Still, the possibility exists that the government can distribute the burdens of military service fairly and justly across the citizenry, perhaps via a random lottery. But even then, the cost of civic participation will be an inefficient allocation of societal resources and perhaps result in a suboptimum labor-to-capital mix for the military since the cost of labor would be artificially suppressed.[6]

In classical liberal thought, on the other hand, emphasis is placed upon protecting the liberty of individuals to pursue their own interests to the extent possible without abridging the liberty of others. As such, the coercive power of the state is to be limited to the minimum necessary to ensure the enforcement of this social contract. Enforcement is necessary to some degree since a public good such as national defense will be undersupplied by the market—that is, by individuals acting in their own self-interest—and this will endanger the polity. Thus coercive taxation, rather than service in kind, is justified to allocate societal resources to

---

1   This argument is different from the variant of subjective control of the military supposed by Stephan Pfaffenzeller, "Conscription and Democracy: The Mythology of Civil-Military Relations," *Armed Forces & Society*, 36:3 (April 2010): 488-9.

2   "[The draft] is far more typical of totalitarian nations than of democratic nations. The theory behind it leads directly to totalitarianism. It is absolutely opposed to the principles of individual liberty which have always been considered a part of American democracy," said Senator Robert Taft (quoted by Harry A. Marmion, *The Case Against a Volunteer Army* (Chicago, IL: Quadrangle Books, 1971), p. 37.

3   Milton Friedman, "Why Not a Voluntary Army?" in *The Draft and Its Enemies: A Documentary History*, edited by John O'Sullivan and Alan M. Meckler (Urbana, IL: University of Illinois Press, 1974), p. 258.

4   Everett C. Dolman, *The Warrior State: How Military Organization Structures Politics* (New York: Palgrave Macmillan, 2004).

5   The preparations for *Ketsu Go*, the Imperial Japanese plan for defense of the home islands against an anticipated American invasion perhaps comes closest to the complete mobilization of the citizenry in time of war. See Richard B. Frank, *Downfall: The End of the Imperial Japanese Empire* (New York: Random House, 1999), pp. 288-330.

6   Pfaffenzeller, "Conscription and Democracy," p. 484.

collective defense.[7] The amount of taxation can be adjusted to make sufficient resources available so that the burdens of military service can be borne by those most willing to bear them, subject to labor market conditions, comparative advantage, and attitudes toward civic duty. It will also likely lead to an optimum labor-to-capital mix for the military given that the true cost of manpower is taken into account.

The downside of the classical liberal solution to national defense is that it results in a degradation of civic life. Although the process of selection for service is just, few provide the public good of defense for the many through their labor as opposed to their taxes. The inequitable distribution of the labor burden leads to a chasm between those who choose to serve and those who do not. This occurs because providing the option of choice enhances the value of service for those who choose it and cheapens the value of citizenship and feelings of ownership in the polity for those who do not—and, historically, they compose the majority. This can lead to the alienation of the military from society and pose problems for the polity.[8]

Over the course of its history the United States has manned its armed forces through both processes. It used conscription to meet its manpower needs in the Civil War, World War I, World War II, the Korean War, and Vietnam. The Vietnam War brought conscription under intense pressure and highlighted general and specific inequities in how the burden of military service was borne. Indeed, as American society gained uncommon prosperity and the distribution of opportunities to share in it became perhaps the most salient issue of the era, "'who serves when most do not serve?' turned out to be those least able to spend the resources to avoid induction—namely, the poor and the black."[9] It was determined that this outcome undermined civic life in the polity and a different approach was required.

In 1973, the United States replaced conscription with an All-Volunteer Force (AVF). The 1970 Presidential Commission on an All-Volunteer Armed Force, chaired by former Secretary of Defense Thomas Gates, provided the rationale for this change and noted five primary concerns of critics:

---

7   Pfaffenzeller, "Conscription and Democracy," p. 493.

8   Gary Schaub, Jr. "Civil-Military Relations," in *The Encyclopedia of Political Science, Volume 1* (Washington, D.C.: CQ Press, 2011); Peter D. Feaver, "The Civil-Military *Problematique*: Huntington, Janowitz, and the Question of Civilian Control," *Armed Forces and Society*, 23:2 (Winter 1996).

9   Richard V.L. Cooper, "Military Manpower Procurement Policy in the 1980s," in *Military Service in the United States*, edited by Brent Scowcroft (Englewood Cliffs, NJ: Prentice-Hall, Inc., 1982), pp. 163-4. Also see Bernard D. Rostker and Curtis L. Gilroy, "The Transition to an All-Volunteer Force: The U.S. Experience," in *Service to Country: Personnel Policy and the Transformation of Western Militaries*, edited by Curtis L. Gilroy and Cindy Williams (Cambridge, MA: MIT Press, 2006).

(1) an all-volunteer force will become isolated from society and threaten civilian control; (2) isolation and alienation will erode civilian respect for the military and hence dilute its quality; (3) an all-volunteer force will be all-black or dominated by servicemen from low-income backgrounds; (4) an all-volunteer force will lead to a decline in patriotism or in popular concern about foreign policy; (5) an all-volunteer force will encourage military adventurism.[10]

The Gates Commission considered these problems and argued that they would be minimal. After nearly four decades of experience, however, concerns remain. Secretary of Defense Robert Gates and Chairman of the Joint Chiefs of Staff Admiral Michael Mullen have discussed the effect that the AVF has had on civic life that echo those considered by the Gates Commission. "Whatever their fond sentiments for men and women in uniform, for most Americans the war remains an abstraction—a distant and unpleasant series of news items that do not affect them personally," said Secretary Gates.[11] "In the absence of a draft, for a growing number of Americans, service in the military, no matter how laudable, has become something for other people to do," he continued.[12] "'With each passing decade, fewer and fewer Americans know someone with military experience in their family or social circle,' Gates said, citing a study showing that the share of 18-year-olds with a veteran parent had fallen from 40 percent in 1988 to 18 percent by 2000."[13]

It is not just the civilian leadership that is concerned. "Adm. Mike Mullen, the chairman of the Joint Chiefs of Staff, told members of the West Point graduating class of 2011 that it was their obligation not only to lead Army units but also to help narrow a widening and worrisome divide between the American public and its military."[14] "'I fear they do not know us,' he said. 'I fear they do not comprehend the full weight of the burden we carry or the price we pay when we return from battle.' He warned that 'a people uninformed about what they are asking the military to endure is a people inevitably unable to fully grasp the scope of the responsibilities our Constitution levies upon them.'"[15]

Gates and Mullen have also indicated and illustrated some negative potential of this estrangement for American civic life. "There is a risk over

---

10   *Report of the President's Commission on an All-Volunteer Armed Force* (Washington: U.S. Government Printing Office, February 1970), p. 129.

11   Anne Flaherty, "Gates Says Too Few in US Bear the Burdens of War," *Associated Press* (September 29, 2010).

12   Elisabeth Bumiller, "Gates Fears Wider Gap between Country and Military," *The New York Times* (September 29, 2010).

13   David Alexander, "U.S. military faces strains after decade of war: Gates," *Reuters* (September 29, 2010). The study is available at: http://prhome.defense.gov/MPP/ACCESSION%20POLICY/PopRep2008/summary/poprepsummary2008.pdf/

14   Thom Shanker, "At West Point, a Focus on Trust," *The New York Times* (May 21, 2011).

15   Thom Shanker, "At West Point, a Focus on Trust," *The New York Times* (May 21, 2011).

time of developing a cadre of military leaders that politically, culturally, and geographically have less and less in common with the people they have sworn to defend," said Secretary Gates.[16] Differentiation between self-selected groups and increasing homogenization of views within those groups is a consequence of group identification and it has had nearly 40 years to materialize.[17] The resultant "gap" in the views of the military and civil society has caught the attention of scholars and has been analyzed extensively over the past 20 years.[18] The biggest fear raised in this literature is that the military could become politicized and use its influence to affect policy just like any other interest group, with dire effects for the polity and the profession.[19]

Indeed, the military could become involved in politics and claim a preferred place in the polity. "On a visit with troops in Iraq Thursday, U.S. Secretary of Defense Robert Gates faced a question on the minds of many of those in uniform: If the federal government shuts down, do the troops still get paid? … 'Well, first of all, let me say you will be paid,' Mr. Gates said, to which some soldiers huddled around the secretary replied, 'Hooah!'—an Army expression of thumbs-up enthusiasm. 'As a historian, it always occurred to me that a smart thing for government was always to pay the guys with guns first,' Mr. Gates jokingly added."[20] Perhaps more seriously, Admiral Mullen echoed the sentiments of these troops. "'We … can't kid ourselves,' he said in a speech last year. 'As much as our young men and women appreciate

---

16   Anne Flaherty, "Gates Says too few in US bear the Burdens of War," *Associated Press* (September 2010).

17   Sue E. Berryman, *Who Serves? The Persistent Myth of the Underclass Army* (Boulder, CO: Westview Press, 1988), p. 55. More generally, see Michael A. Hogg, *The Social Psychology of Group Cohesiveness: From Attraction to Social Identity* (Washington Square, NY: New York University Press, 1992) and Peter J. Burke and Jan E. Stets, *Identity Theory* (New York: Oxford University Press, 2009).

18   Peter D. Feaver and Richard Kohn (eds), *Soldiers and Civilians: The Civil-Military Gap and American National Security* (Cambridge, MA: MIT Press, 2001); Feaver, Peter D. and Christopher Gelpi, *Choosing Your Battles: American Civil-Military Relations and the Use of Force* (Princeton, NJ: Princeton University Press, 2004); and Christopher Gelpi, Peter D. Feaver, and Jason Reifler, *Paying the Human Costs of War: American Public Opinion and Casualties in Military Conflicts* (Princeton, NJ: Princeton University Press, 2009).

19   Richard H. Kohn, "How Democracies Control the Military," in *The Global Divergences of Democracies,* edited by Larry Diamond and Marc F. Plattner (Baltimore, MD: The Johns Hopkins University Press, 2001); Richard H. Kohn, "The Erosion of Civilian Control of the Military in the United States Today," *Naval War College Review,* 55:3 (Summer 2002); Richard H. Kohn, "The Danger of Militarization in the Endless 'War' on Terrorism," *The Journal of Military History,* 73:1 (January 2009); and Richard H. Kohn, "Building Trust: Civil-Military Behaviors for Effective National Security," in *American Civil-Military Relations: The Soldier and the State in a New Era,* edited by Suzanne C. Nielsen and Don M. Snider (Baltimore, MD: Johns Hopkins University Press, 2009).

20   Nathan Hodge, "U.S. to pay 'Guys with Guns' First," *Wall Street Journal* (April 7, 2011).

the gestures of kindness we see today in tribute to our military and our veterans, a free ticket to a football game or a pat on the back will not solve their problems.'... Gates, Mullen and others usually conclude their warnings about the civil-military divide by listing ways they hope Americans will help the one million or so troops who have served in Iraq and Afghanistan. For example, Mullen wants government and private-sector employers to give preference, or at least an equal shot, to veterans looking for jobs. He has also warned that Americans must be prepared for decades' worth of support to keep today's veterans from becoming homeless in the numbers that the military saw after Vietnam."[21] Even as the United States has suffered from the worst economic conditions since the Great Depression and the largest deficits in its history, the military budget has been considered nearly sacrosanct.[22] Indeed, it has been argued that it is very difficult for political leaders to undertake any policy that does not appear to "support the troops," and that this degradation of civic life is an unintended consequence of adopting the AVF.[23]

Given these concerns, we assess the following propositions:

1. Is the AVF demographically representative of American society in terms of race and ethnicity, gender, and geographic origin?
2. Is the AVF ideologically representative of American society in terms of party identification and ideological affinity?

Who Joins? Accessions

When the United States shifted to an AVF, the military became just another employer in the eyes of many, especially labor economists.[24] Each year, the military must

---

21   Philip Ewing, "Defense Leaders fear Gap Between Public, Military," *Politico. com* (February 20, 2011). The speech referenced was delivered to the American Security Council Foundation on October 27 2010, http://www.ascfusa.org/content_pp./view/adm-mullens-ausa-speech/

22   See, for instance, Thom Shanker, "Gates Takes Aim at Pentagon Spending," *The New York Times* (May 8, 2010); Thom Shanker, "Pentagon Told to Save Billions for Use in War," *The New York Times* (June 3, 2010); Thom Shanker and Christopher Drew, "Pentagon Faces Growing Pressures to Trim Budget," *The New York Times* (July 22, 2010); Ginger Thompson and Thom Shanker, "Generals Wary of Pentagon Move to Cut Their Ranks," *The New York Times* (August 27, 2010); Elisabeth Bumiller and Thom Shanker, "Gates Seeking to Contain Military Health Costs," *The New York Times* (November 29, 2010); John T. Bennett, "Military Spending Balloons Amid Bipartisan Calls for Cuts," *TheHill. com* (May 31, 2011).

23   Andrew J. Bacevich, *The New American Militarism: How Americans are Seduced by War* (Cambridge: Oxford University Press, 2005).

24   Colin Ash, Bernard Udis, and Robert F. McNown, "Enlistments in the All-Volunteer Force: A Military Personnel Supply Model and Its Forecasts," *American Economic Review* 73:1 (March 1983). Concern for this effect on the military as an

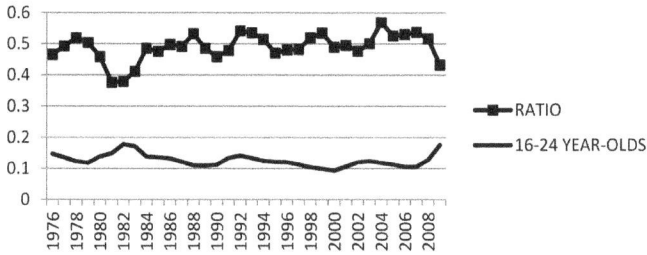

Figure 7.1   Enlisted Accessions: Applicant Acceptance and 16-24 Year Old Unemployment Rates

Source: Office of the Under Secretary of Defense for Personnel and Readiness, *Population Representation in the Military Services* (Washington, D.C.: Department of Defense, 2009), Table D-3. Ratio of Non-Prior Service (NPS) Active Component Enlisted Accessions to Applicants, FYs 1976-2009 and Table D-2. Annual Civilian Unemployment Rate by Age Group, FYs 1973-2009.

induct enough members to meet its desired end strength, which requires it to replace those who separate or retire. This accession process is the entry point to the force and therefore is an important step in shaping its representativeness.

Enlisted Accessions

In the debate over the transition to an AVF, it was argued that poor blacks would be more likely to enlist in the military than would be expected given their representation in the population.[25] It was also argued that enlistment standards—education, aptitude, conduct, and physical fitness—would check this potential surge. It is the case that the AVF has been selective in accepting applicants to the armed forces: it accepts roughly 50 percent of applicants. The acceptance rate does vary in response to other factors: the authorized end strength of the armed forces, retirements, separations, and reenlistment rates determine demand for accessions while the size of the relevant population and the state of the economy, in particular civilian employment opportunities for the population likely to enlist (that is, 18-24 year olds with a high-school education), determine supply. Figure 7.1 graphs the acceptance rate of applicants to the enlisted ranks from 1976 to 2009. As can be seen, the rate generally hovers around 50 percent with decreases occurring during the recessions of 1979-

---

institution was expressed by Charles C. Moskos Jr. "From Institution to Occupation: Trends in Military Organization," *Armed Forces and Society*, 4:1 (Fall 1977).

   25   Gates, *Report of the President's Commission on an All-Volunteer Armed Force*, pp. 143-53; Friedman, "Why Not a Volunteer Army?" p. 260; Armor and Gilroy, "Changing Minority Representation," p. 227.

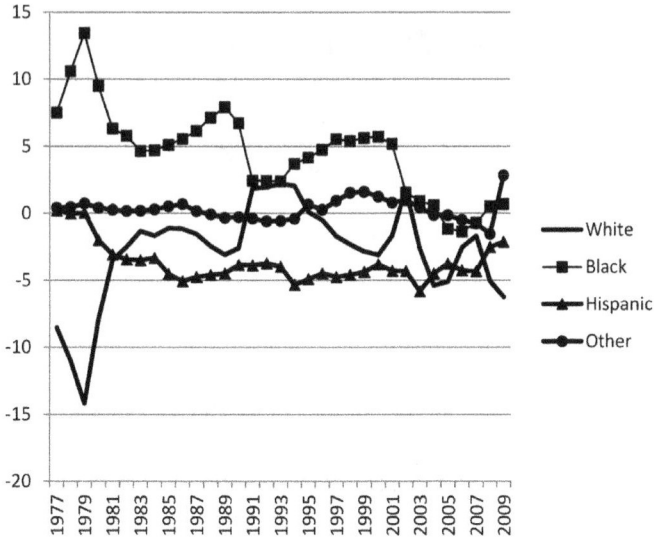

Figure 7.2   NPS Accession Differences by Race/Ethnicity

Source: Office of the Under Secretary of Defense for Personnel and Readiness, *Population Representation in the Military Services* (Washington, D.C.: Department of Defense, 2009), Table D-22. Non-Prior Service (NPS) Active Component Enlisted Accessions by Race with Civilian Comparison Group, FYs 2003-2009, Table D-23. Non-Prior Service (NPS) Active Component Enlisted Accessions by Race/Ethnicity, FY 1973-2002, and Table D-24. Non-Prior Service (NPS) Active Component Enlisted Accessions by Ethnicity with Civilian Comparison Group, FYs 2003-2009.

1983 and 2008-2010 with their associated jump in unemployment for 16-24 year olds[26] and the increased during the worst years of the Iraq War.

Were minorities relatively more likely to enlist into the military than whites given their proportion in the population? Figure 7.2 charts how over- or under-represented different racial and ethnic groups were in terms of non-prior service enlisted accessions from 1977 to 2009. It compares the proportion of accessions by race and ethnicity with each group's share of the 18-24 year old non-institutionalized civilian population. As can be seen, during the early years of the AVF blacks were over-represented by seven to 13 percent and whites under-represented by roughly the same amount. Armor and Gilroy attribute this to an error in scoring the military aptitude test that inflated scores at the

---

26   The Pearson product moment correlation coefficient, r, indicates a -0.667 relationship between the unemployment rate of 16-24-year-olds and the enlistment acceptance rate.

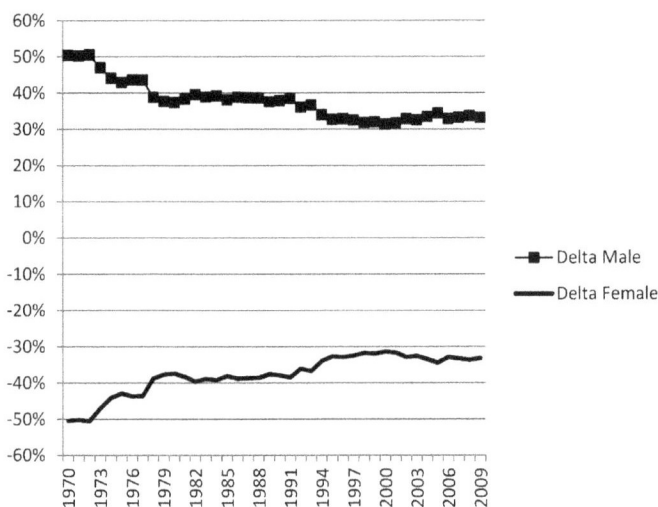

Figure 7.3   Percentage of Non-Prior Service Enlisted Accessions: Gender Differential with Civilian Population

Source: Office of the Under Secretary of Defense for Personnel and Readiness, *Population Representation in the Military Services* (Washington: Department of Defense, various years).

lower end of the spectrum.[27] Accession ratios tended to normalize thereafter, albeit with whites under-represented (except during 1991-1995 and 2002) and blacks over-represented (except during 2005-2007). Hispanics have tended to be under-represented by about five percent throughout the period. Only the residual category of "other" minorities (for example, Asians, Native Hawaiians/ Pacific Islanders, American Indians/Alaska Natives, and more than one race or ethnicity) accesses into the military consistently in proportion to their numbers in the population.

Turning to gender, military service has traditionally been seen as a masculine endeavor[28] and only in recent times have women been integrated into most, but not yet all, aspects of the military.[29] Figure 7.3 presents a comparison of male and female enlisted accessions compared with their proportions in the 18-24

---

27   Armor and Gilroy, "Changing Minority Representation," p. 229 and p. 245, fn. 35.

28   See, for instance, James H. Webb, Jr., "The War on Military Culture," *The Weekly Standard* (January 20, 1997) and Adam B. Lowther, "The Post-9/11 American Serviceman," *Joint Force Quarterly* 58:3 (Fall 2010), p. 77.

29   See Martha McSally, "Women in Combat: Is the Current Policy Obsolete?" *Duke Journal of Gender Law and Policy*, 14 (2007) for an overview.

year old non-institutionalized civilian population from 1970-2009.[30] As we can see, women have been greatly under-represented among those enlisting into the armed forces, ranging from 50 percent in 1970 to 32 percent in 2009. But there has been a distinct trend toward greater female representation among enlisted accessions, which perhaps portends greater changes to the masculine nature of military service and civic life more generally.

It is the case that American culture has become increasingly homogenized as advances in transportation and communications technology interact with its social and economic mobility, yet regional differences remain and affect civic life.[31] Secretary Gates has argued that youths from certain regions of the country are more likely to enlist in the military. Indeed, as seen in Figure 7.4, accessions from the South have been greater than all other regions throughout the entire period of the AVF—and they have increased significantly since 1985. Likewise, accessions from the West have increased throughout the period and now are second. Accessions have declined from the Midwest and Northeast regions of the country, although the former stabilized and have slightly increased since 1993. Still, the report upon which Secretary Gates based his claim admitted that these trends "could be due to shifts in the civilian population over time" but did not evaluate this possibility.[32] Comparison with a relevant civilian cohort is required to make any judgments. Figure 7.4 compares regional accession data to high school graduates without a college degree in 2003, 2004, 2008, and 2009.[33] As can be seen, the discrepancies are not as large as they seemed without the comparison. High school graduates without a college degree in the South do tend to be over-represented in enlisted accessions by six-seven percent and in the West by roughly five percent, while those from the Midwest and Northeast tend to be under-represented by three and six percent, respectively. Therefore, while Secretary Gates is correct that there is a recruiting bias in the American military, it is not as pronounced as some have claimed.[34]

---

30   Office of the Under Secretary of Defense for Personnel and Readiness, *Population Representation in the Military Services* (Washington: Department of Defense, various years).

31   Michael Morgan, "Television and the Erosion of Regional Diversity," *Journal of Broadcasting & Electronic Media*, 30:2 (1986); Robert W. Jackman and Ross A. Miller, "A Renaissance of Political Culture?" *American Journal of Political Science*, 40:3 (August 1996); Joseph A. Vandello and Dov Cohen, "Patterns of Individualism and Collectivism Across the United States," *Journal of Personality and Social Psychology*, 77:2 (August 1999).

32   Center for Naval Analyses, *Population Representation in the Military Services: Fiscal Year 2008 Report prepared for the Undersecretary of Defense (Accession Policy), Summary* (Alexandria: Center for Naval Analyses, 2008), p. 35.

33   Regrettably, these statistics were not reported regularly.

34   Gary Schmitt and Cheryl Miller, "The Military Should Mirror the Nation," *The Wall Street Journal* (August 26, 2010), p. 15.

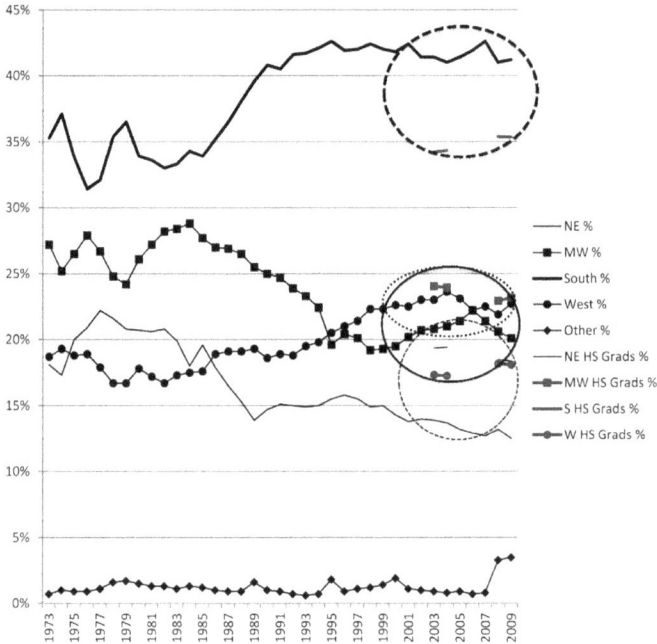

Figure 7.4   Non-Prior Service (NPS) Active Component Enlisted Accessions by Geographical Region with Civilian Comparison, FYs 1973-2009

Source: Bureau of Labor Statistics, *Geographic Profile of Employment and Unemployment* (Washington, D.C.: Department of Labor, various years).

Officer Accessions

The adoption of the AVF was not expected to affect the officer corps, which had been a volunteer professional force for roughly a century.[35] The Gates Commission argued that "[o]fficers will continue to be recruited from all over the nation and from a variety of socio-economic backgrounds."[36] Despite this characterization, the officer corps has always been more representative of the advantaged classes in society, requiring a college degree for entry. Therefore one could presume that whites would be over-represented in the officer corps and minorities under-represented. Indeed, as one critic of the AVF argued, "[w] hatever else can be said about the officer corps of a volunteer system, it would be

---

35   Russel F. Weigley, "The American Military and the Principle of Civilian Control from McClellan to Powell," *The Journal of Military History*, 57:5 (October 1993).

36   Gates, *Report of the President's Commission on an All-Volunteer Armed Force*, p 138.

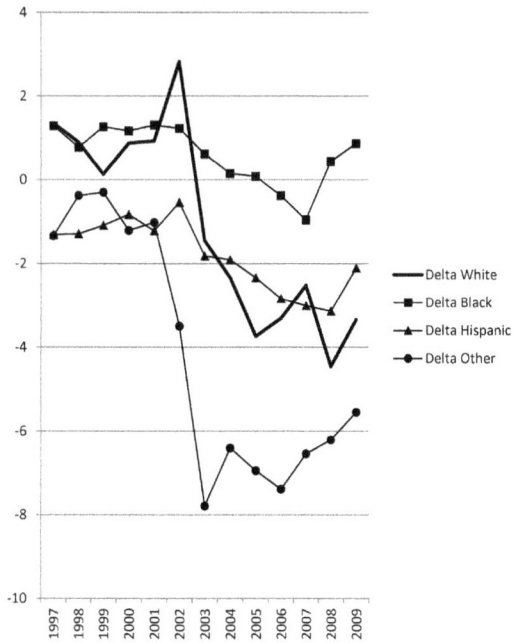

Figure 7.5   Officer Accession Differential by Race/Ethnicity

Source: Office of the Under Secretary of Defense for Personnel and Readiness, *Population Representation in the Military Services* (Washington, D.C.: Department of Defense, various years).

lily-white."[37] Figure 7.5 presents the differential for officer accessions compared with 21-35 year old civilian college graduates in the non-institutional civilian population from 1997-2009. As can be seen, whites were over-represented in officer accessions from 1997-2003 and under-represented since. Blacks were over-represented throughout except for the 2006-2007 period. Hispanics and "other" minorities have been most under-represented in officer accessions throughout the period.

As with enlisted accessions, gender representation in officer accessions is substantially biased toward males, with males over-represented by roughly 35 percent and females under-represented by the same amount. And, as shown in Figure 7.6, this has been relatively constant since 1997—which also matches the enlisted accession trend during this period.

---

37   Harrt A. Marmion, *The Case Against a Volunteer Army* (Chicago, IL: Quadrangle Books, 1971), p. 52.

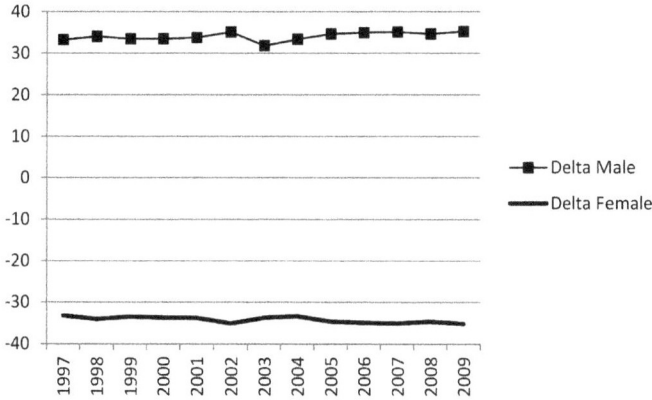

Figure 7.6   Officer Accession Differential by Gender

Source: Office of the Under Secretary of Defense for Personnel and Readiness, *Population Representation in the Military Services* (Washington, D.C.: Department of Defense, various years).

Who Serves? The Force

Accession data informs us with regard to who joins the military in any given year but it does not tell us who serves. This can be quite a different matter as changes in accessions may take a decade or more to have a significant effect on the force as a whole. Below we examine the profile of the active duty force.

The Enlisted Force

We saw above that in the period 1997-2009, "other" minorities joined the military in proportion to their numbers in the relevant civilian population, black accessions declined from roughly five percent above to 1 percent below before stabilizing around zero differential, Hispanics were chronically under-represented by five percent, and white accessions varied significantly but were generally under-represented. How did these trends affect the racial and ethnic composition of the enlisted force?

Figure 7.7 shows that blacks are over-represented in the enlisted force as compared to their 18-24-year-old counterparts in the civilian workforce. Yet their proportion has declined gently this past decade from 10 percent to roughly 6 percent more than we would otherwise expect. Hispanics remain chronically under-represented in the enlisted ranks by 4.5-6 percent. Whites are also under-represented by roughly 7 percent from 1997-2002 and 10.5-13 percent thereafter. Finally, "other" minorities

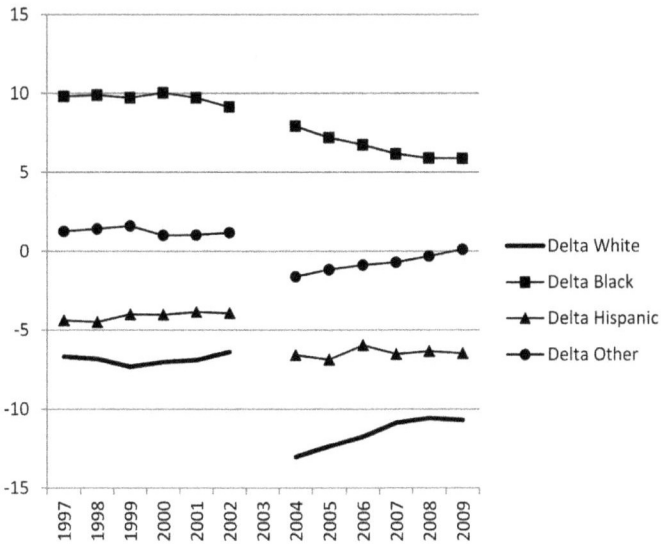

Figure 7.7   Enlisted Force Differential by Race/Ethnicity

Source: Office of the Under Secretary of Defense for Personnel and Readiness, *Population Representation in the Military Services* (Washington, D.C.: Department of Defense, various years).

are included in the enlisted ranks in rough proportion to their counterparts in the civilian world.

The story with regard to gender is roughly the same as it was with accessions: males are consistently over-represented by over 30 percent and females are consistently under-represented by a similar amount, which given the 35 percent accession rate bias, suggests that females are more likely to remain in the force longer than males.

The Officer Corps

During the 1997-2009 period the military commissioned proportionally more black officers than one would expect, proportionally more whites until 2003, and proportionally fewer Hispanics and "other" minorities. How has this reflected itself in the officer corps? As shown in Figure 7.9, the over-representation of white officers relative to the civilian population of 21-49 year old college graduates in the civilian workforce dropped precipitously from 1997 to 2003 and declined slightly thereafter, resulting in under-representation since 2006. Black officers are represented in the officer corps in rough proportion to their counterparts in

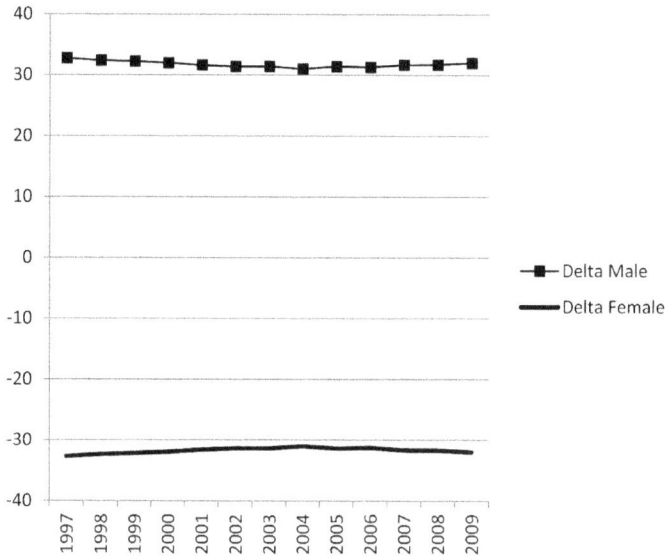

Figure 7.8   Enlisted Force Differential by Gender

Source: Office of the Under Secretary of Defense for Personnel and Readiness, *Population Representation in the Military Services* (Washington, D.C.: Department of Defense, various years).

the civilian population.[38] Hispanic officers are under-represented as are "other" minority officers, whose proportions declined significantly after 2003.[39]

In terms of gender, Figure 7.10 shows that the stable rate of difference in officer accessions is reflected almost perfectly in the officer corps, with males over-represented by 33-34 percent relative to their college educated civilian counterparts and females under-represented by 33-34 percent.

Where do officers come from? The Gates Commission argued that "Officers will continue to be recruited from all over the nation and from a variety of socio-economic backgrounds."[40] This appears to hold true. Although the Department of Defense regularly reports figures for the commissioning source of officers—

---

38   Still, the representation of black officers throughout the ranks continues to be an issue. See Irving Smith III, "Why Black Officers Still Fail," *Parameters* 40:3 (Autumn 2010) and Gary N. Leong, "Diversity Within the U.S. Air Force Senior Leadership," (Professional Studies Paper, Air War College, February 17 2010).

39   This could be an artifact of the manner in which DoD coded this category. In 2003 it disaggregated it into AIAN, Asian, NHPI, and Two or More.

40   Gates, *Report of the President's Commission on an All-Volunteer Armed Force*, p. 138.

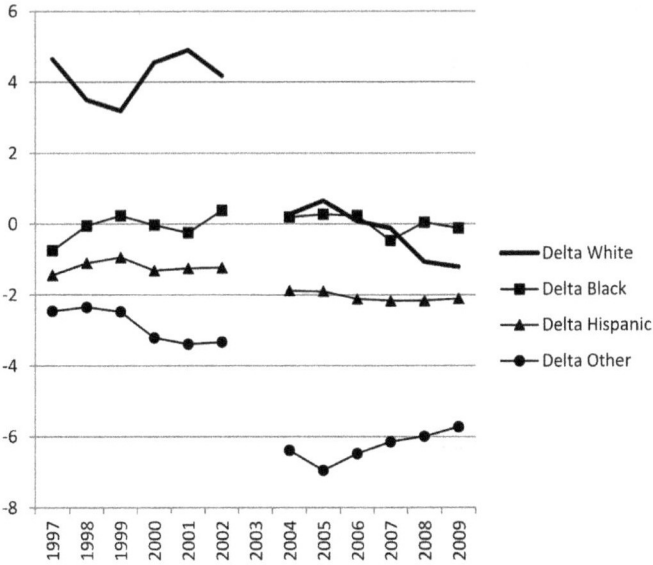

Figure 7.9   Percentage of Officer Corps Differential by Race/Ethnicity

Source: Office of the Under Secretary of Defense for Personnel and Readiness, *Population Representation in the Military Services* (Washington, D.C.: Department of Defense, various years).

through military academies, ROTC, OTS, and direct appointment—it does not report the geographic origins of the officer corps.[41] However, some data is available to make a tentative assessment of where this part of the force originated. Figures 7.11[42] and 7.12 compare the region of origin of elite officers who participated in the Triangle Institute for Security Studies survey of Feaver and Kohn while enrolled in in-residence professional military education in 1998-1999[43] and in the Officer Strategic Leadership Survey while enrolled in in-residence professional military education in 2006-2010.[44] While these two

---

41   It records, but does not report, the home of record of all members.

42   These figures were too busy when combined.

43   See Peter D. Feaver, Richard H. Kohn, and Lindsay P. Cohn, "The Gap Between Military and Civilian in the United States in Perspective," *Soldiers and Civilians: The Civil-Military Gap and American National Security*, edited by Peter D. Feaver and Richard Kohn (Cambridge, MA: MIT Press, 2001) for information about their sample.

44   The Officer Strategic Leadership Survey was administered to officers attending the School of Advanced Air and Space Studies and the Air War College in February 14-March 13, 2006. Officers attending those schools and Air Command and Staff College were surveyed in May 2007, August 20-September 10, 2008, October 1-16, 2009, and September

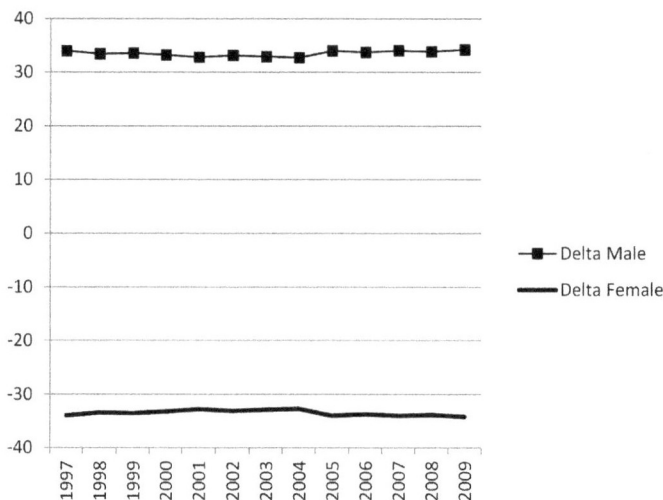

Figure 7.10   Officer Corps Differential by Gender

Source: Office of the Under Secretary of Defense for Personnel and Readiness, *Population Representation in the Military Services* (Washington, D.C.: Department of Defense, various years).

samples vary in terms of service balance, reflecting the school houses where the surveys were administered, they constitute the best time-series data available.[45]

Secretary Gates' concern about the force being increasingly biased toward Americans hailing from the South and Mountain West[46] is not borne out in these samples when comparable civilian populations are considered.[47] As seen in Figure 7.11, although surveyed officers hailing from the South have increased from roughly 20 percent to 28 percent from 1998 to 2010, they are still under-

---

2010. Officers from the Naval War College were surveyed in June 2007. In all, 1104 officers were surveyed.

45   Urben's "Civil-Military Relations in a Time of War Survey" (2009) captured this information for 3,901 Army officers in 2009. We considered combining it with the 2009 OSLS data, but decided against it so as to not skew the consistent population from which that data was drawn. See Heidi A. Urben, "Civil-Military Relations in a Time of War: Party, Politics, and the Profession of Arms," (Ph.D. dissertation, Georgetown University, April 14, 2010).

46   Neil Offen, "Defense Secretary Makes Service Pitch," *Durham Herald-Sun* (September 30, 2010), p. 1.

47   Bureau of Labor Statistics, *Geographic Profile of Employment and Unemployment* (Washington, D.C.: Department of Labor, various years). Regrettably, these statistics were not regularly reported.

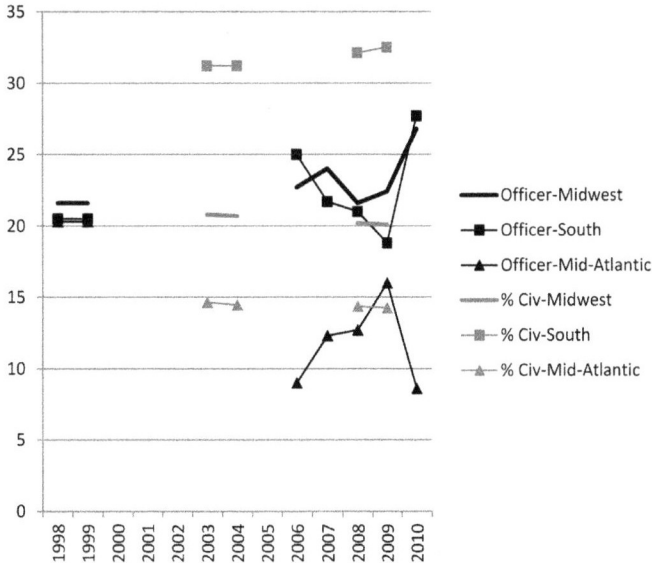

Figure 7.11   Percentage of Officers by Region: TISS and OSLS 1

represented compared to their numbers in the college-educated civilian labor force by five or more percent. On the other hand, the equally numerous surveyed officers with origins in the Midwest are over-represented when compared to college-educated civilians by two to seven percent. Officers from the Mountain states represent the smallest regional contingent in the officer corps and are under-represented when compared to their civilian counterparts, as are officers coming from the Pacific states (Figure 7.12). Concerns about the under-representation of officers hailing from New England seem to be misplaced, as surveyed officers from there constituted 11 percent of the surveyed force in 1998-1999 and six to eight percent between 2006-2010—which is greater than their proportions among the comparable civilians, which is around six percent.

Finally, what about the political and ideological diversity of the officer corps? "[C]ritics of the AVF anticipated that voluntary entry would produce a military with values increasingly distinct from civilian values...Free choice of military service was therefore expected to increase the homogeneity—and therefore distinctiveness—of values within the military. Moreover, it was expected to reduce military turnover, and AVF critics implicitly knew what social psychology repeatedly shows. The longer a person is a member of a group, the more socialized that individual becomes to the values of the group

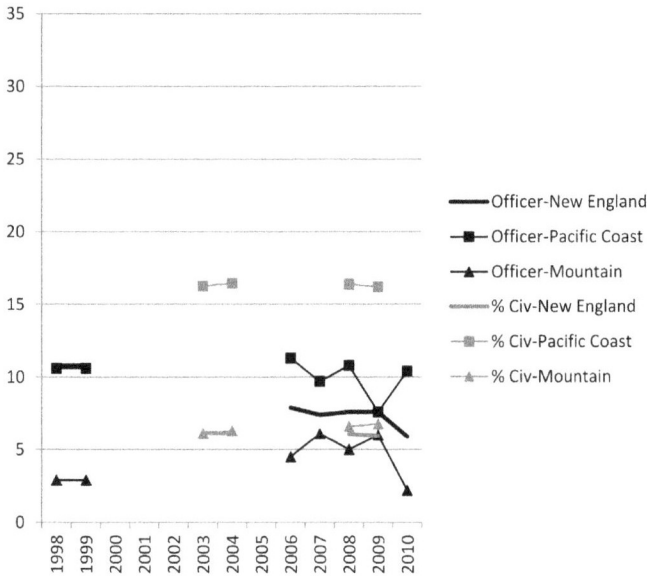

Figure 7.12　Percentage Officers by Region: TISS and OSLS 2

and the more similar that person's values become to those of the group."[48] Have these fears been borne out with regard to political attitudes?

It would seem so. Numerous studies have demonstrated that a majority of the officer corps self-identify with the Republican Party and are ideologically conservative.[49] The data presented in Figures 7.13 and 7.14, gathered by the Officer Strategic Leadership Survey for 2007-2010 and compared with the TISS data of 1998, add to the consensus. The consistency across the samples is striking, with 67-70 percent of officers identifying as conservative and 57-74 percent self-identifying with the Republican Party. This might not be exceptional if such identifications were reflected in the civilian population, but officers self-identify with the Republican Party and with conservative political

48　Sue E. Berryman, *Who Serves? The Persistent Myth of the Underclass Army* (Boulder, CO: Westview Press, 1988), p. 55.

49　Ole R. Holsti, "A Widening Gap Between the U.S. Military and Civilian Society? Some Evidence, 1976-96," *International Security*, 23:3 (Winter 1998/1999); Ole R. Holsti, "Of Chasms and Convergences: Attitudes and Beliefs of Civilians and Military Elites at the Start of a New Millennium," in *Soldiers and Civilians: The Civil-Military Gap and American National Security*, edited by Peter D. Feaver and Richard Kohn (Cambridge, MA: MIT Press, 2001); Jason K. Dempsey, *Our Army: Soldiers, Politics, and American Civil-Military Relations* (Princeton, NJ: Princeton University Press, 2010); and Urben, "Civil-Military Relations in a Time of War."

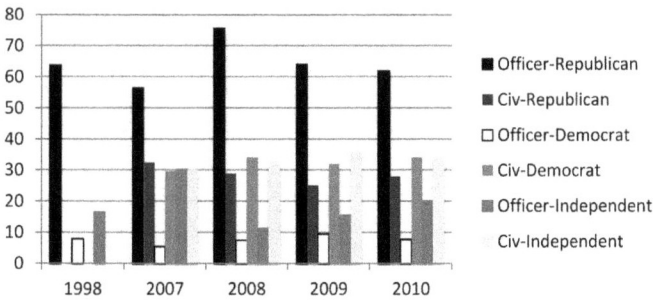

Figure 7.13 Percentage of Party Identification: Officers and Civilian College and Graduates

ideology far more than comparable civilians with a college education or better.[50] As can be seen in Figure 7.13, officers are more Republican (+24-47 percent), less Democratic (−23-4 percent), and even less Independent (+0-22 percent) than comparable civilians. In this regard, they are clearly unrepresentative of the American polity. Officers are also quite different ideologically, being far less liberal (−11-17 percent), far more conservative (+30-32 percent), and less moderate (−15-21 percent)—as seen in Figure 7.14.

This trend would seem to be a function of socialization and self-selection. Urben found that one out of five Army officers "admitted to becoming more conservative since joining the Army," that "officers who affiliated with the Democratic Party were more likely to express discomfort talking about politics in the workplace," and that "junior officers who were leaving Army service affiliated with the Democratic Party at a higher rate than the rest of their junior officer peers." On the other hand, she "found little evidence to suggest that service in the Army causes its officers to increasingly affiliate with the Republican Party" or that "service-specific variables affect officers' affiliation with the Republican Party."[51]

The Chairman of the Joint Chiefs of Staff has repeatedly emphasized the apolitical nature of a professional officer corps. "Admiral Mullen said that to maintain the public's trust, the new officers must remain apolitical. He described their role as 'a neutral instrument of the state, accountable to our civilian leaders, no matter which political party holds sway.'"[52] "'Keeping our politics private is a good first step,' he added. 'The only things we should be wearing on our sleeves are our military

---

50   Civilian data derived from Pew surveys of June 2007, September 2008, October 2009, and October 2010. Strata indicated are "College graduate (B.S., B.A., or other 4-year degree)" and "Post-graduate training or professional schooling after college (e.g., toward a master's Degree or Ph.D.; law or medical)."

51   Urben, "Civil-Military Relations in a Time of War," pp. 152-3.

52   Thom Shanker, "At West Point, a Focus on Trust," *The New York Times* (May 21, 2011).

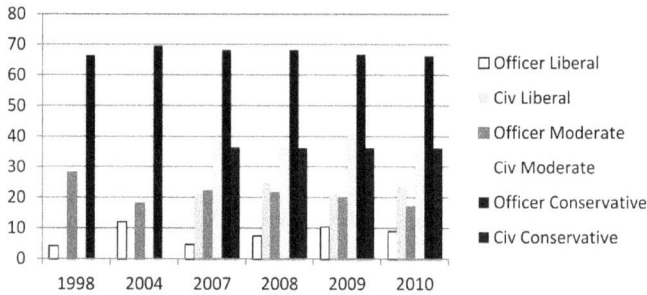

Figure 7.14 Political Ideology: Officers and College and Civilians

Source: 2004 officer ideological self-identification data taken from the Citizenship and Service Survey of Dempsey for lieutenant colonels and colonels, as reported by Urben, "Civil-Military Relations in a Time of War," p. 45, Table 2.7.

insignia.'"[53] It seems clear that to the extent that partisanship and politicking have not been a cause of alarm among the political classes, it is because the professional identity of officers as nonpartisan and apolitical servants of the polity has been working overtime to keep their personal political predilections in check.[54]

Conclusions

The United States has been at war since September 11, 2001. It has conducted major operations in Afghanistan and Iraq and minor ones in the Philippines, Pakistan, Yemen, Somalia, Libya, and many other exotic lands. Over 6,044 American service men and women have been killed and 43,643 wounded as of Memorial Day 2011.[55] Despite a decade of war, "the all-volunteer uniformed services now represent less than one percent of the American population, but

---

53   Quoted by Thom Shanker, "Top-Ranking Officer Warns U.S. Military to Stay Out of Politics," *The New York Times* (May 25, 2008). Also see Leo Shane III, "Mullen Reminds Military Leaders to Stay Professional, Apolitical," *Stars and Stripes* (January 10, 2011).

54   Urben "found that most Army officers voted at a rate greater than the general public, [but] also found other measures of political activity to be generally muted for Army officers," (Urben, "Civil-Military Relations in a Time of War," p. 153).

55   Department of Defense, "Operation Iraqi Freedom (OIF) U.S. Casualty Status, Fatalities as of: May 31, 2011, 10 a.m. EDT; Operation New Dawn (OND) U.S. Casualty Status, Fatalities as of: May 31, 2011, 10 a.m. EDT; Operation Enduring Freedom (OEF) U.S. Casualty Status, Fatalities as of: May 31, 2011, 10 a.m. EDT," http://www.defense.gov/news/casualty.pdf

they're carrying 100 percent of the battle."[56] The strains of the current wars have brought attention to this unbalanced burden, yet it is the result of the attempt to fix the unbalanced burden from the last long American War: Vietnam.

When the United States government opted for an AVF in 1973 it did so fully aware that it was choosing a just process over a just outcome. Over the course of a generation, this means of manning the force resulted in an enlisted force that became more balanced in terms of race, ethnicity, and gender. Today's wars are not being fought by "the poor and black." It has also resulted in an officer corps that has also become more demographically balanced over time: it is not "lily-white." Indeed, in terms of race, ethnicity, and gender, the focus has shifted from general representation within the force—especially the officer corps—to representation across all ranks and especially in the leadership.[57] As such, diversity concerns have gone beyond President Truman's Executive Order 9981 that called for "equality of treatment and opportunity for all persons in the armed services."[58]

Concerns about the AVF went beyond demographic representation, however. The Gates Commission considered the possibility that "an all-volunteer force will become isolated from society and threaten civilian control."[59] They concluded that "the change to an all-volunteer force will have no effect on top leadership, since these men have always been professionals."[60]

Yet being a military professional does not preclude officers from holding political views or affiliating with a political party—General George Marshall's example notwithstanding. "We must be apolitical and I think we've watched that erode over time, over the last 20, 30, 40 years," said Admiral Mullen.[61] The literature and the data presented here supports the Chairman's view: officers overwhelmingly self-identify as Republicans, they are over seven to ten times more likely to than to identify with the Democratic Party, and twice as likely to do so than civilians with comparable educational attainment. It also indicates that officers are six to 10 times more likely to be politically conservative than liberal, two to three times more likely to identify as conservative than moderate,

---

56   Tom Brokaw, "The Wars that America Forgot About," *The New York Times* (October 17, 2010).

57   Military Leadership Diversity Commission, *From Representation to Inclusion: Diversity Leadership for the 21st Century Military. Final Report* (Arlington, TX: Military Leadership Diversity Commission, March 15, 2011).

58   Military Leadership Diversity Commission, *From Representation to Inclusion*, p. vii.

59   Gates, *Report of the President's Commission on an All-Volunteer Armed Force*, p. 129.

60   Gates, *Report of the President's Commission on an All-Volunteer Armed Force*, p. 138.

61   Quoted by Greg Jaffe, "Adm. Mike Mullen Observes Disconnect Between U.S. Military and Broader Public," *The Washington Post* (January 11, 2011).

and twice as likely to be conservative than comparably educated civilians. The persistence of these trends over time suggests that the officer corps is becoming quite unreflective of the polity.

But does this matter? Can political views affect civilian control of the military in the United States? It seems unlikely, but the classical liberal strain of conservative political ideology is distrustful of governmental authority and these views may find their way into the officer corps. Indeed, Urben finds in her sample of 4,248 Army officers that "Republican-leaning officers were more likely to display lower trust levels in the government compared to Democrats. For example, 62 percent of Republicans felt that when civilians gave orders to the military, domestic partisan politics were the motivation compared to 53 percent of Democrats. Forty-one percent of Republicans believed that during wartime civilians should let the military run the war, compared to 31 percent of Democrats. And with regards to the question on respect for the Commander-in-Chief, 46 percent of Republicans felt the president should have served in the military, compared to just 18 percent of Democrats."[62] Republican-leaning officers are also "more likely to advocate military leaders being assertive and insistent when offering military advice on the use of force."[63] As she concludes, "it is not an encouraging sign."[64] It also suggests that the degradation of civic life that has animated the recent concerns of Secretary Gates and Admiral Mullen has affected the military as well. And in this, perhaps, the military all too much representative of the society from which it is drawn.

---

62 Urben, "Civil-Military Relations in a Time of War," p. 124.
63 Urben, "Civil-Military Relations in a Time of War," p. 138.
64 Urben, "Civil-Military Relations in a Time of War," p. 139.

Chapter 8

Military Theory, Strategy and Praxis: Implications for Civil-Military Relations

Jacob W. Kipp and Lester W. Grau

[The opinions expressed are those of the authors and do not necessarily represent those of the Department of Defense or U.S. Government.]

> I want to speak to you tonight about our effort in Afghanistan—the nature of our commitment there, the scope of our interests, and the strategy that my administration will pursue to bring this war to a successful conclusion. (President Barack Obama, West Point, New York, December 1 2009)[1]

This chapter was first written in March 2010 and began with the quote from the address by President Barack Obama cited above in which he addressed U.S. strategy in Afghanistan. His remarks remain a good point of departure for a discussion of national strategy and can now be evaluated in light of our experience in Afghanistan and the very recent remarks of the President on June 22, 2011, which addressed the way forward in Afghanistan, including a timetable to withdraw the surge forces from Afghanistan and a deadline of 2014 for a shift from combat to support operations, when "he Afghan people will be responsible for their own security." This course of action, the President described as "the beginning—but not the end—of our effort to wind down this war."[2] That change is the product of the progress made on the ground against the Taliban as a result of the surge announced in 2009 and the successful joint special operation, "Operation Neptune Spear," which led to the elimination of the leader of al Qaeda, Osama bin Laden, at his compound in Abbottabad, Pakistan, on May 2, 2011.[3]

---

1   This essay originally appeared as a slightly different article under this title in *Military Review* (March-April 2011):12-22. The authors express their appreciation to the editor of *Military Review* for granting permission for its use here.

2   "Remarks by the President on the Way Forward in Afghanistan," The White House, June 22, 2011, http://www.whitehouse.gov/the-press-office/2011/06/22/remarks-president-way-forward-afghanistan. Accessed June 28, 2011.

3   "Behind the Hunt for Bin Laden," *The New York Times*, May 2, 2011, http://www.nytimes.com/2011/05/03/world/asia/03intel.html; and "Osama Bin Laden Dead: Raiders Knew Mission A One-Shot Deal," *Huffpost World*, May 17, 2011, http://www.

Success hinges upon the management of risk over time in a whole-of-government effort to prepare the Afghan state, its military, police, other institutions, and economy to take on the task of achieving a settlement that will protect its population and provide stability for the nation's development. According to the President, U.S. policy will seek to find elements among the Afghan resistance with which to negotiate as its prosecutes counterterrorism operations against al Qaeda and its allies among the Taliban. This course of action depends very much on the response of elements with the Taliban resistance, which, while united by Islamic fundamentalism, are divided by tribal sentiments and their commitment to the use of terror against fellow Muslims. The domestic context of the President's recent remarks on Afghanistan and the timeline for that conflict make transparent the political content of his message, as does the current debate between the executive branch and congress over the nature of military operations undertaken "to protect the civilian population of Libya" by the United States and its allies under a NATO mandate, which has raised the question of when combat operations become a war requiring the President to ask for congressional approval under the War Powers Resolution.[4] In Afghanistan, the President has stated that he continues to act according to the vital interests of the United State. But over Libya, American forces are, according to the President, engaged in limited air operations to achieve humanitarian ends within the context of a civil war in that country. If we invoke the test of praxis over time and admit the possibility that air strikes may not achieve their "humanitarian objectives," then one would be right to ask what then follows: abandonment of those objectives or escalation to resolve the conflict? In either case prudent questions at this point demand strategic answers. However, that admission conceals a glaring problem. Strategy today is not what it was during the Cold War or even World War II. There is a radical difference between strategy formulated to fight conventional wars and deter nuclear wars and that necessary to conduct armed struggle in the post-modern world. The state no longer defines the nature of the conflict in the latter case.

A review of the literature on war and military thought quickly reveals an interesting bias. The authors often cited are those of the Western military tradition with a few ancient authors, one or two Chinese, and a passing reference to some Russian or Soviet thinker or practitioners.[5] The military theoreticians of old still hold sway in the staff colleges and war colleges of the world's professional militaries. Western students have at least a nodding acquaintance with the writings of Clausewitz, Jomini, Du Picq, Douhet, Fuller, Liddell-Hart, Machiavelli, Mahan and Upton. Interested students also investigate Sun Tsu. Advanced students study

---

huffingtonpost.com/2011/05/17/osama-bin-laden-raid-one-shot-deal_n_862900.html. Accessed June 28, 2011.

4   "Obama: 'Fuss' over Libya action is just 'politics'''," *CBS News*, June 29, 2011, http://www.cbsnews.com/8301-503544_162-20075425-503544.html. Accessed June 29, 2011.

5   Martin van Creveld, *The Art of War: War and Military Thought* (New York: Smithsonian Books, 2005).

Svechin, Triandafilov and Tuchachesky to appreciate operational art. Professionals need to know the foundations of their profession and much of the old theory is still applicable. Over the last decade, in the face of the challenges posed by terrorism and insurgency, a larger community of officers has returned to counterinsurgency and low-intensity conflict and even named it another generation of war, the fourth. Of course, Mao, Lawrence, Giap and Galula are still read. But contemporary authors addressing the complexity of counterinsurgency have gained currency. These include Martin van Creveld, William Lind, Joe Celeski, Shimon Naveh, and David Killcullen. They will also know the works of John Boyd, Deitrich Doerner, Arthur Cebrowski, and William Owens.

Unlike the earlier theory of warfare, which was based on the nations-at-war model, which emphasized the primacy of conflicts conducted between nations and saw the constabulary functions of countering brigands and pirates as a necessary but secondary task, contemporary theory has to give a central place to combating non-state actors. Since 2001, with the exception of a few weeks in the spring of 2003, the United States and its allies have been making war on non-state actors, quasi-organizations beyond the brigand or pirate status, but clearly not state actors. Indeed, the territory of these non-state actors encompasses that of several states yet they formally control little of that territory. And although the agents of these non-state actors impose their control over local judicial systems and religious practices, the actors carry out few functions of a state. Their very persistence reinforces the impression that in some parts of the world the Western conception of the nation-state born with the Treaty of Westphalia is under challenge.

This different sort of conflict is challenging the way armed forces organize, equip, and conduct themselves in the face of this threat. When the U.S. Army and Marine Corps issued FM 3-24, the introduction described that publication as filling "a doctrinal gap" relating to the conduct of counterinsurgency operations. The experience associated with operations in Iraq and Afghanistan drove the doctrine writers. But as the manual makes clear, the political dimension of the counterinsurgency demands strategic adaptation as well as tactical and operational adjustments. It is a matter for the whole of government and not just the military in the field.

These conflicts radically redefined how Western states address such conflicts and incorporate other elements of national power. If a decade ago, staff colleges taught DIME (diplomatic, informational, military, and economic) as the elements of national power and students sought to apply MIDE (military, informational, diplomatic and economic) power to their staff problems, today discussions of conflict begin with complexity theory, systems analysis and design. Before one can plan a campaign, one must understand the problem at hand. Today's problems defy templating.[6] While Army discussions of design have focused on operational art, logic demands that

---

6   Huba Wass de Czege. "The Logic of Operational Art: How to Design Sound Campaign Strategies, Learn Effectively and Adapt Rapidly & Appropriately," unpublished essay, January 2009, p. 2.

design be applied to strategy since it is the point in the process where the political dimension should first be addressed.[7] Naveh, Challans and Schneider have called this reorientation "the structure of operational revolution."[8] And, while this is a revolution, it is also a negation of the autonomy of operational art precisely because it imposes the centrality of strategy at the highest level by imposing political direction at the start and retaining control of political intervention throughout the campaign and during subsequent campaigns as reframing the conflict becomes necessary. In this context, the informational element involves the development of a narrative that has the power to explain actions taken and contemplated.[9] It has strategic impact in that the information element feeds directly into the political process.

The past few decades have seen a fundamental shift in the application of technology to warfare. It has brought about serious changes in the organization of military institutions. The informatization of conventional warfare made possible the conduct of network-centric warfare and precision strikes across the depth of the battlefield demolishing purely force-on-force correlations-of-forces models and introducing a new calculus (and modeling) based upon computational power, networks, sensors, and guidance systems. This new technology has had a profound impact on tactics, organization and funding priorities of those possessing such capabilities and of those potentially facing such capabilities. The dialectical struggle between the sides has meant no clear winner. On some occasions, advanced technology has brought the achievement of profound successes for those empowered by such technology. On other occasions, those lacking advanced technologies have shown an ability to adapt to the threats posed by such systems and engage in protracted struggles which democracies find hard to sustain.[10] Immediate operations in Afghanistan in the fall of 2001 brought lightning success against Taliban field forces and seemed to confirm the decisive impact of Transformation on the conduct of war. But then, the appearance of a post-Saddam insurgency in

---

7   The Army discussion of design has been associated with the annual war game run by JFCOM at the Army War College for the Army Chief of Staff under the title "Unified Quest." These discussions led to the publication of TRADOC Pamphlet, 525-5-500 "Commander's Appreciation and Campaign Design" in January 2008, which addresses an approach to problem framing and design before commanders actually begin operational planning under MDMP (military decision making process). In this context, design implies a political-military dialog among political leaders and military commanders before planning, during planning, during execution and following execution.

8   Shimon Naveh, Jim Schneider, and Timothy Challans, *The Structure of Operational Revolution* (Washington, D.C.: Booz Allen Hamilton, 2009).

9   In the work which gave first prominence to the term "operational art" [*operativnoe iskusstvo*], Aleksandr Svechin spoke of the risk of adopting a strategy of annihilation, which in practice meant transforming all problems into matters of operational art and solvable by combat power and reduced politics to a secondary role. See Aleksandr A. Svechin, *Strategy* (Minneapolis, MN: Eastview Press, 1992), p. 240.

10   Lester Grau and Jacob Kipp, "The Fog and Friction of Technology," *Military Review* (September-October 2001), 5, 88-97.

Iraq and the reconstitution of the Taliban in Afghanistan and Pakistan forced major adjustments to the conduct of both wars. In retrospect, both insurgencies may have been preventable or less-severe through proper planning, proper resourcing and finishing what was started. An insurgency is always weakest at the beginning and is not, by definition, protracted. The insurgent strives to make it so to offset government power.

The bias of modern militaries and their political leaderships toward seeking decision by annihilation has resulted in frustration when confronted by a protracted struggle. In such cases war is not the continuation of politics. Rather war assumes a political content all its own, which, in fact, reshapes the content of war itself. That insight is not that new. Clausewitz, who took part in the campaign of 1812 as a member of the Russian staff, saw first-hand how political content could frustrate military genius by invoking the concept of people's war. In 1812, Napoleon lost in Russia without a single decisive defeat. Swarms of partisans, winter, and the dogged pursuit of the Russian Army embodied what Lev Tolstoy called *narodnaia voina* (people's war). Indeed, in early 1813, while still serving as a Russian officer, Clausewitz took part in the effort to ignite a popular war against France in Prussia, while his King was still an ally of Napoleon.

But Clausewitz discussed this problem in the context of a Newtonian universe of laws. Now military theorists have to confront a world of quantum mechanics and complex systems generating wicked problems. Good planning cannot overcome a fundamental misunderstanding of the problem that military power has been called upon to address. Decision by annihilation gives way to protracted struggle, where the very advantages of advanced technology seem negated. Technology, which seemed to liberate warfare from the risk of stalemate on the battlefield, now seems unmanned by the complexity of war among the peoples. Meanwhile, the military educational institutions that once taught Clausewitz as the chief theorist of modern war, have had to relook the problem of "small wars" and insurgency. War theory lags and technology is no substitute for theory.

Under Transformation, as practiced by the Department of Defense under Secretary Donald Rumsfeld, technology became a substitute for theory since it was assumed that the U.S. military would use its informational advantage and network organization to quickly defeat any opponent in the field and deter most from engaging in conflict. Two protracted wars later, this assumption is proven wrong. The unstated assumption of the technological determinists was that a simple template could be applied to each and every conflict, where technology would leverage a rapid and decisive outcome. In the aftermath of Desert Storm, Operation Deliberate Force, Operation Allied Force, and Operation Enduring Freedom, it seemed that such was the case. There were messy details—the survival of Saddam, the protracted deployments into Bosnia Herzegovina, the negotiated end of NATO's war over Kosovo, and the survival of remnants of al Qaeda and the Taliban. But they were not enough to stimulate a profound debate about ends, ways and means. Instead, when planning turned to Iraq, the issue was the size of the force needed to achieve rapid decision against the Iraqi army in the field and the speedy occupation of Baghdad. The post-conflict environment was simply assumed to be benign, permitting the rapid redeployment of U.S. and Allied forces out of Iraq.

But insurgencies are like Tolstoy's unhappy families, they are each unique and, as such, demand complex study to understand their dynamics. This is, of course, almost impossible for the intervening power before it applies force, since it lacks the deep expertise about the society in question. However, the longer the war continues, the more apparent it becomes that such expertise is necessary to define the political center of gravity of the conflict as well as the allegiances of the various groups that make up the population. Nation-building, which assumes that an ersatz model of Western institutions and values can be imposed on these populations, fundamentally misses the point. Stability will come when the population in all its complexity has grounds to assume that its security is assured. No checklist of projects, which the occupier assumes represents the wishes of the population, will serve as a reliable guide to progress. That can only be determined by feedback from the local population and the task of collecting such feedback will never be easy in a foreign land of a distinctly different culture with an armed insurgency under way.

Soldiers are not likely to be the best agents for collecting such information, whether they are foreign troops or national troops unconnected to the local population. Home guard units and local police can serve to provide such information, but their primary loyalty will be to the immediate security of their community. Building trust with them takes time and great efforts. It means accepting the protracted struggle, which the insurgents see as their road to victory. Short of making the effort to understand the desires of the local populations, armies will be tempted to apply a template of violence to intimidate the insurgents and accept "collateral damage" to non-combatants as a necessary cost on the road to military victory. That such damage actually broadens the base of the insurgency and makes both the national government and the occupying force seem to be hostile oppressors does not become apparent until after the damage has been done. The point is to apply violence in a directed fashion against enemy combatants, much in the manner that a constabulary applies violence in the protection of the community it is supposed defend against lawless actions. For soldiers on the ground, this demands a much different set of rules of engagement than those practiced in high-intensity conflict and comes much closer to that applied in the cases of the domestic application of martial law. These new situations demand a clear rethinking of strategic priorities.

Strategy addresses the ends, ways and means of war and embraces how a nation prepares and conducts war. There are essentially four components to strategy: the economic, political, military, and informational.[11] Strategy determines how the

---

11 The dual content of informational power in this formulation often goes uncommented upon. In systemic terms it means the information generated about friendly and hostile ends, ways, and means and the engagement in strategic communication to create a narrative that explains national choices and counters enemy information operations. Of course, this involves many elements of national intelligence. But it demands a convincing

state will fight the war, what the state seeks to be the phases of that war and under what conditions and how the state will terminate the conflict. Strategy sets ends, ways and means so that political and military leaders can determine progress, or lack of progress, in implementing a strategy.[12] Leaders, however, must explain their conduct to their citizens, the larger international community, and last but not least the population directly affected by the conflict. This implies both knowledge of the population in question and the existence of means to solicit feedback from that population over the course of the conflict. Close examination of most theaters of conflict reveals the existence of many communities within each population that must be monitored. This last point is an admission that this population cannot be treated as "the other" or as unfortunate complications to a neat battlefield where firepower can be applied without constraints against the enemy force in the field. In this sense strategy recasts the conduct of operations and the application of tactics. It is an admission that soft power may be more effective in achieving stability than kinetic means.

Strategic assessment helps determine how successful various courses of action might be, and once the conflict has begun permits the review of the course of the conflict and the likelihood of success in following a particular strategy. For eight years the United States and its allies have been directly involved in the Afghan conflict without a comprehensive strategy. Our initial intervention was punitive, designed to punish al Qaeda and the Taliban for protecting al Qaeda. Half-hearted efforts at state-building followed while Washington shifted its attention to Iraq. In the meantime al Qaeda survived, the Taliban recovered and became a source of armed insurgency in both Afghanistan and Pakistan. Even though counterinsurgency experts agree that the solution to a guerrilla conflict lies primarily in the political and economic realm, down to 2009, no systematic exposition of national or alliance strategy was forthcoming.

In the last year, President Obama stated that the Afghan conflict was a necessary conflict and recast it to embrace both Afghanistan and Pakistan. President Obama's speech at West Point outlined the first clear attempt at his articulation of a national strategy in Afghanistan. Down to this point, the struggle in Afghanistan appeared to be an open-ended commitment to the application of military power in a protracted

national narrative to explain a course of action, the costs, and the outcome. Implausible narratives rather quickly collapse in the face of facts on the ground, or as, Stalin used to say, "Facts are stubborn things." National policy based upon finding and destroying weapons of mass destruction when none could be found comes to mind as a telling example. But one could also look at the conflict in Afghanistan that is described as a fight with the Taliban, when the armed resistance is much more diverse and the conflict more complex.

12   This is not the U.S. official view. The U.S. definition of strategy is "A prudent idea or set of ideas for employing the instruments of national power in a synchronized and integrated fashion to achieve theater, national, and/or multinational objectives." Joint Publication 1-02. This definition may be part of the problem. Strategy is so much more than a prudent idea.

war, in which success was both undefined and remote and depended most upon the continued application of limited though growing combat power. Strategy seemed to be in the hands of the generals without a political dimension (which makes it a military strategy but hardly an over-arching national strategy). After a long review in consultation with his political and military advisors, President Obama articulated a strategy for Afghanistan. Critics may argue over the size of the additional deployment, the chances of success on the ground, and even the importance of the conflict in determining national priorities. But a strategy has been articulated for conflict that is assumed to be necessary to U.S. and NATO interests. One should not confuse the articulation of a strategy with a prediction of the course and outcome of the conflict. There are too many variables beyond the power of even the United States to control. In the final analysis it will be the peoples of Afghanistan and Pakistan who will determine the outcome of the conflict. The most that American and NATO intervention can achieve is to assist in setting the conditions for a settlement among Afghans that will enhance regional stability and reduce the threat of terrorist attacks emanating from Afghan and Pakistani territory.

Time will tell whether the current strategy incorporated the right elements to manage the conflict to a successful conclusion—a settlement among Afghans that will enhance regional stability and reduce the threat of terrorist attacks emanating from Afghan and Pakistani territory. Every strategy's chances of success depend upon getting the correct definition of the problem in order to apply elements of national power to its solution. Strategy is dialectical in that success depends upon the enemy's responses in the struggle for the loyalty of the population. And this is not a macro problem subject to a grand exercise in templating. It depends upon local dynamics, which require deep knowledge of each region and its population, which becomes an exercise in understanding the human terrain and plotting its evolving features.

Recent wars have uncovered a glaring national strategic weakness—the inability to plan beyond a military ends, ways and means mission. This weakness has been exacerbated by the changing nature of warfare conducted by U.S. opponents. National strategic thinking and planning is running behind its advancing military without the proper integration and employment of assets. Lessons are not being learned as the drawn-out nature of U.S. conflicts demonstrate.

How Did the Mismatch Occur?

During World War II, military theory, strategy and praxis were in balance. The Cold War and Korean Conflict operated both within and outside comprehensive strategy, since the assumption was that nuclear exchange would destroy the planet, so strategy involved the prevention of nuclear exchange and emphasized the growth of the military component and military technology at the expense of the political and economic. Conventional maneuver war was to be conducted at the operational level under nuclear-threat. The nuclear balance of terror dominated relations and restrained risk. The antagonists poked at each other using proxies in limited contests in South Vietnam, Angola, Afghanistan and numerous "Wars of National Liberation." With the collapse of the Soviet Union, the bipolar nature of global relations dissolved. The West was in ascendency, but how would theory, strategy and praxis adapt to the new reality? Did nuclear terror matter in a world without nuclear stand-off, but with the remaining "superpower" unwilling or unable to lead the planet in other than the conventional military dimension? What would be the impact of regional nuclear proliferation, which put nuclear weapons into the hands of states disposed to national conflicts along ethnic and religious lines? At the same time, the much-heralded economic dominance of the remaining "superpower" has faded as it became a debtor nation with a much-smaller industrial base and the source of credit excesses that shocked global financial markets.

Desert Storm—The Stage Setter

Operation Desert Storm set the stage for the current dilemma. The lesson that the potential opponents of American power took away from this operation was quite evident: trying to match the technologically advanced ground, air and naval forces of the United States was a sure path to military, if not political, defeat. The U.S. Armed Forces had trained to take on the Soviet Union and, given a half-year to prepare the theater, were unbeatable in Kuwait against a less formidable foe, which had fought the Iranians to a stalemate in the 1980s. The only apparent way left to oppose the U.S. Armed Forces and its allies was to adapt Liddell-Hart's strategy of "the indirect approach" to the twenty-first century and mitigate the technological overmatch that the U.S. depended upon for speedy victory. This involved moving the contest to an area where that technology would be degraded—forest, jungle, mountains, delta or urban centers and make military mass disappear, replacing regular formations with guerrillas and partisans. This is the point that William Lind made in his articles on fourth generation warfare. It was the subtext to all the discussions of "asymmetric warfare" in the 1990s.

Kosovo

The Serbs provided the first post-Desert Storm conflict for NATO and U.S. Armed Forces in Kosovo. The Serbs learned from the Iraq experience that camouflage was effective for the Iraqis and moved their army into the mountains and forests, hid their systems and turned the engines off. They built dummy mockups of tanks, bridges and command posts. Their goal was to preserve the army for post-conflict use. They were successful. The planned three-day air operation lasted 78 days. The Serbs did not surrender but negotiated a settlement via the EU on terms better than those initially offered by NATO. The air forces had accurately destroyed its target set, which included real military facilities and a good share of mockups and, when that did not bring about Serbian defeat, turned to the civilian infrastructure as the primary target with air strikes destroying power plants, transportation nodes, and bridges, which disrupted commerce in the Danube region for years. West Germany, Russia and Finland finally intervened and negotiated a settlement that left the Serbian government intact, postponed the issue of Kosovo's independence and resulted in a long-term occupation mission for NATO.

The Serbian Army emerged from the woods. Trained analysts counted the battalion sets as the units drove out. They were mostly intact. The Serbian Army survived. The John Warden's concentric-circle adaptation of Douhet's theory of air power reduced civilian casualties but it could not impose a political defeat on an opponent who still held the ground in contention. Kosovo ended with a negotiated settlement, when it appeared that NATO would have to risk fracture over the combat deployment of ground troops into Kosovo. The Clinton administration's narrative of victory through airpower alone began to disintegrate and threaten alliance solidarity. In spite of this, the air-only operation was acclaimed by some as the new face of warfare. Future war would involve U.S. air power and would be supplemented by somebody else's ground forces. There was no need for U.S. ground forces in future conflicts. They would arrive as part of an allied occupying force to serve as a constabulary to maintain a settlement dictated by air strikes. Significantly, this view of future war still did not incorporate a system for conflict termination beyond continuous bombing and assumed no economic or political costs for the side mounting the air offensive. Any delay in war termination was simply a matter of adjusting the target set to achieve the right physical and psychological destruction against the targeted actor, which in this case was not the nation but its political and military elite.

Afghanistan

Afghanistan provided the second post-Desert Storm conflict. The United States had been attacked. The strategic narrative was one of a punitive expedition to punish those who launched those attacks. The Bush administration, especially Secretary of Defense Donald Rumsfeld, wanted to recreate Desert Storm with the sophisticated technology that a decade of acquisitions had provided. But Afghanistan was not Kuwait or Iraq and none of the conditions of Desert Storm applied. It was not a prepared theater. The United States did not have a half-year to prepare while moving massive stocks and forces into position along the border. The United States did not want to commit its own ground forces. It wanted another Kosovo with U.S. airpower and someone else's army defeating the Taliban and al Qaeda. Although Afghanistan was nominally a state, the Taliban was mostly a government in name only—a government of a failing or failed state.

Based on advice from Pakistan, the United States wanted to replace the Pashtun-Taliban with a Pashtun government drawn from the Durrani tribal group—the traditional rulers of Afghanistan. The United States needed a Pashtun force to defeat a Pashtun force. Further, the Pashtun force needed to support a Durrani government, yet the Durrani were the power base of the Taliban. The majority Pashtun tribal group, the Ghilzai, had their own ambitions and goals. The United States enlisted the help of an old friend, Abdul Haq, to raise a Pashtun force to fight a Pashtun force. The United States had already launched an air operation against Afghanistan. It was an air operation designed against a prepared theater targeting the Taliban integrated air defense system, Taliban integrated command and control system, Taliban tank maintenance facilities and Taliban logistics columns. None of these "target sets" made much sense against the Taliban and it was clearly not a prepared theater. The air operation quickly ran out of targets and then rearranged rubble and killed civilians. Abdul Haq, trying to recruit his Pashtun force, begged that the air operation cease because of the civilian casualties it created and because the targets struck were of little advantage in defeating the Taliban and al Qaeda, but his pleas were ignored. The only real target in the country was the Taliban and al Qaeda field forces deployed against the Tajiks, Uzbeks, Hazara (and some Pashtun) who belonged to the so-called Northern Alliance. The Taliban and al Qaeda were a conventional force, deployed in a linear fashion. With good ground spotters, they were an optimum target for air strikes. They were deployed in a single echelon, had no meaningful reserves and no national mobilization capacity, thus making the field force a very fragile target. Initially, this target was ignored. The United States, for political reasons, did not want the Northern Alliance to break out and seize the country.

Then, on October 25, 2001, Abdul Haq was killed by the Taliban. There would be no Pashtun force to defeat a Pashtun force. Without committing U.S. ground forces, the Northern Alliance was the only available force. US Special Operations teams had joined the Northern Alliance forces. They could provide effective ground observation and adjustment to air strikes. When the forces of the Northern Alliance, U.S. airpower and special operations combined, they quickly overcame the Taliban and al Qaeda forces deployed in static positions. The Taliban and al Qaeda pushed out rear guards, abandoned the cities and went to the mountains. After the initial shock, the enemy retreat was coherent and it succeeded in preserving its leadership, its logistics structure and much of its force. The U.S. effort did not have a plan or the capability to complete the defeat of the enemy and run the country. The United States assumed that it had won since it now controlled the cities. This was the same mistake that the Soviets and British had made. It soon became obvious that al Qaeda and the Taliban represented movements that could rally political support and raise irregular forces to fight an insurgency. In the meantime time, the United States introduced conventional ground forces, which were able to smash the remaining conventional enemy forces. But there still was no long-term strategy for dealing with the Pashtun problem or establishing a post-conflict order in Afghanistan.

During this interval it would have been useful for U.S. political and military leaders to have a deep understanding of Afghanistan, and particularly the pattern of warfare practiced in Afghanistan. This pattern starts with the defeat of Afghan conventional forces, and then devolves into a low-grade marginally-effective guerrilla war. The occupier hardly knows that there is a guerrilla conflict going on and is more concerned with criminality than combating a guerrilla. Over time, the overly bold and stupid are eliminated from the guerrilla force and it becomes more competent and able to challenge the new government and the occupying force. The force does not evolve into a regular army and even risks defeat if it tries to challenge the occupying force in conventional battles. One day the new government and the occupier are faced with a full-blown insurgent threat. The guerilla force tries to win over the countryside and gradually strangle the cities.[13]

Iraq

The invasion of Iraq was the third post-Desert Storm conflict. It was clear that someone else's army was not available to overthrow Saddam Hussein. It was a prepared theater with coalition logistics bases well established, LOCs in good repair and forces positioned forward. The coalition had ample time to get set and into position (although Turkey's intransigence prevented getting forces in place for an initial northern axis). When the invasion occurred, some Iraqi camouflage measures succeeded, but it is difficult to hide everything in an open desert. SCUD missiles are one thing, divisions are another. The Armed Forces of Iraq resisted effectively in some areas, but in others, they felt it was useless to fight, so they went home. Shortly after the invasion, two FMSO analysts went to Iraq and interviewed Iraqi military personnel. Their story was universal—"the officers left and I went home." The Fedayeen resistance, however, was prepared to engage the United States in guerrilla warfare. They had trained for it—and they were equipped.

US airpower had proven very effective against the Iraqi conventional forces and had been very constrained in attacking civilian targets. One result was the lack of wide-spread damage to Baghdad and other cities. The air forces were very precise in their targeting and left most of the infrastructure intact. This precision and concern for the civilian population may have actually worked to the coalition disadvantage. When talking to Iraqi civilians, several of them commented that "were we really defeated? Nothing is destroyed. Our army just quit." Baghdad was the anti-Dresden and constrained bombing certainly did not break the will of the civilian populace. Most of them were glad to be rid of Saddam, but many were determined to make the occupier bleed through guerrilla war.

---

13   An occupier can change this calculus by removing the label of occupation from the equation via withdrawal under conditions that strengthen the capacity of the government to practice the traditional Afghan strategy of dividing the opposition and securing its base in the cities. Such an end is not neat, does not involve military victory, and can often depend upon making alliances of convenience with local war lords, tribal leaders, and ethnic communities. Lester W. Grau, "Breaking Contact Without Creating Chaos: The Soviet Withdrawal from Afghanistan," *The Journal of Slavic Military Studies*, Volume 20, April-June 2007, 234-61 and Makhmut Akhmetovich Gareev, *Moya poslednyaya voyna: Afganistan bez Sovetskikh voysk* [*My Last War: Afghanistan without Soviet Forces*], Moscow: INSAN, 1996.

The Way Ahead

The United States Armed Forces were prepared to fight World War III. They were not so ready to fight in the forest, jungle, mountains, delta or urban centers— or to fight guerrillas. The post-conflict stage (phase IV) eluded implementation. The beauty of Mahan, Clausewitz, Douhet and Mao is that they incorporated the political and economic element as part of war theory. Today military planners are searching for "an immaculate victory with arms-length use of cruise missiles, predator drones and special ops."[14] What do you do after you have bounced the Taliban out of position and out of the cities? How do you deal with non-state combatants? How does the civil population fit into the military calculations?

The post-Cold War lesson for the U.S. seems to be that the political and economic areas are vital to post-conflict resolution and have to be an inherent part of strategy. They need to be incorporated in military planning and military theory. War planning should not embrace annihilation at the expense of political calculation and adjustment in the course of the campaign. But political risk aversion should not outweigh coherent, realistic war planning. One can become enamored with Moltke the Elder and his victory at Sedan and miss the point that it was Bismarck who came up with the political strategy that kept France divided between left and right and isolated Paris. The political and economic element needs to be part of that planning. The planning and conduct of a conflict should be in the hands of an integrated national leadership which can discuss among itself the political, economic, and military dimensions of the conflict in a common language in keeping with a democratic polity and an open society.

Predictability, despite all the technological determinists claims, is not to be found in warfare.[15] Embarking or not embarking on a conflict involves risk. The best the national leadership can do is to assess that risk and develop strategy that will minimize it. If embarking upon a conflict involves significant risk, which the society will not accept, the nation ought not to embark upon war in the first place. War has become much more than the continuation of politics by other means. It is at its heart a political process of great complexity in an environment fraught with chaos, which most of its actors understand imperfectly. Understanding a particular war is a labor of Sisyphus, a necessary, difficult and frustrating task defying efforts to impose meaning, unity and clarity upon events. Such efforts are inevitably negated by the evolution of the conflict itself thanks to the interactions of the contesting sides and other actors. War becomes a chameleon changing its appearance and even content before one's eyes. This does not negate the need for theory. Without theory there can be no sound political course of action or strategy.

---

14 Charles Krauthammer, "Afghan War Forces Obama to Make a Real Decision," *Kansas City Star* (October 13, 2009), A13.

15 Antoine Bousquet, *The Scientific Way for Warfare: Order and Chaos on the Battlefields of Modernity* (New York: Columbia University Press, 2009), pp. 242-3.

The immediate task that praxis has put before theory is the need to deal with conflict on difficult terrain—topographical and human. The great guerrilla theorists, Mao Tse-Tung, Lawrence, and Giap, recognized this problem. However, their theories do not apply to Afghanistan because, again, insurgencies are like Tolstoy's unhappy families, each unique in their own environment. This is not the first time a modern force faced a tribal irregular force. The Indian Wars of the United States and the colonial wars of the European powers and the United States come to mind. The Russian and Soviet experiences in Central Asia and the Caucasus also are relevant. But in all these cases, the regular force had as its objective, through punitive expeditions or direct conquest, to incorporate the territory into its domains. Afghanistan may have begun as a punitive expedition but failure to properly finish the job and the resulting political commitments and the revived insurgency make it a particularly difficult problem involving a strategy of attrition and political negotiations.

Strategy is the domain of governments, not just the military, but the political authorities have abandoned strategy making it a military-only concern. The military, in turn, is heavily involved in planning, but strategy is something more. Reducing strategy to being the task of the senior military commander in country and not the government as a whole leads to a military and geographic articulation of choices. This quickly frustrates all effort to formulate a whole-of-government approach to strategy. But any strategy for a particular conflict has wider and deeper implications at home and abroad. Ultimately, it falls to the head of state to explain a strategy, to mobilize the whole of government and to gain and sustain public support in spite of the costs in blood and treasure which will have to be borne. Behind this problem stands the need for shared discourse about national security issues so that the real alternatives can be part of an informed public debate. In the United States, the "bully pulpit" still belongs to the President. These considerations should direct the formation of U.S. strategy toward Afghanistan and Pakistan. Readers of different political persuasions can read President Obama's November 2009 address in different ways, depending upon their own assumptions. But there can be no doubt that the President did articulate a three-part whole-of-government strategy for the United States and its NATO allies to apply to the conflict in Afghanistan and Pakistan.

Praxis and technology can strongly influence but cannot drive theory and strategy. The military situation facing the world today is different, challenging and requiring new approaches, organizations, priorities and theory. The conflicts in Afghanistan and Pakistan do not lend themselves to maneuver warfare, air-centric warfare or effects-based operations but each are relevant to the task of developing a theory of post-modern conflict.[16] The informationization of warfare will go forward and will bring in its wake weapons systems based on new physical principles. Still, changes in military technology and their application

---

16   Still, a leaflet left on an Afghan door promising death to the inhabitants if they cooperate with coalition forces is an effects-based operation.

will not negate the capacity of an adaptive opponent to seek to impose his own strategy and tactics upon a conflict, which he assumes involves his vital interests. This fact alone drives the need for relevant theory and comprehensive strategy that goes beyond the military dimension. The enemy will always have a vote. Praxis attempts to make it an insignificant one. Theory and strategy should be about the ends, ways and means to counter that enemy and incorporate the means to adapt to his changes. Praxis should direct future strategic choices and technology should enable those choices to enhance the conduct of political and military conflict.

Chapter 9

Business Models and Emerging U.S. Warfighting Concepts

Milan Vego

The U.S. military has used for a long time various business practices in managing its bureaucracy, budget, and in force planning. The business management models have repeatedly proven their high value in running the Pentagon. However, during the 1960s the Pentagon extensively used business model in its conduct of war in Vietnam. This experience proved to be utterly disastrous. Yet since the late 1990s, the U.S. military increasingly embraced the notion that business models can be applied to the conduct of war. In doing so, it confused the ends, the means and the ways of business and warfare. Instead of focusing on leadership, the U.S. military emphasis was increasingly put on management, military efficiency instead effectiveness, and application of various quantifiable methods called "metrics" based on the business model in assessing the performance of military forces in combat. Another problem in the U.S. military is increasing use of various business terms in describing purely military activities.

The Roots

During World War II, both the United States and United Kingdom extensively used various business statistical methods for the analysis of the effects of their strategic bombing. They also used various operations research techniques for the analysis of the anti-submarine warfare (ASW) in the Atlantic and offensive mining in the European waters and in the Pacific. In the late 1950s, the U.S. Navy developed a network model called PERT (Program Evaluation and Review Technique) for managing the work of thousands of contractors in its highly successful POLARIS missile program. PERT provided managers a graphical display of the various activities, and estimate of how long each activity and the entire program will take to complete, and which activities are the most important to ensure a timely completion of the program. PERT proved to be a highly successful tool for planning, coordinating, and controlling large complex military programs.

McNamara's Era

A major effort to introduce various business models in the U.S. military came during the tenure of Robert S. McNamara as the Secretary of Defense (1961-1968). The main reason for adopting business practices was McNamara's almost exclusive focus on improving efficiency of the U.S. military. Planning-Programming-Budgeting System (PPBS) became the heart of McNamara's management method of producing long-term defense budget. He was a firm believer that the computational techniques, which had cracked codes in World War II and built jet fighters could be applied across all aspects of business and politics. McNamara also introduced a "game theory" approach to war in Vietnam at the political strategic level. The U.S. would send messages to the enemy, whose responses could then be predicted. He also uses various metrics such as body counts to measure the progress of war in Vietnam. All this had predictable and catastrophic consequences for the U.S. military.[1] McNamara also extensively applied systems analysis run by civilian "whiz kids" as a basis for making key decisions on force requirements, and in designing weapon systems.

Revival of the Business Model

After the early 1970s, McNamara's methods of applying business practices were discredited in the U.S. military. However, in the late 1990s, Secretary of Defense William Cohen directed Pentagon to take advantage of the so-called "revolution in business affairs" to improve efficiency and cut waste. The U.S. military also adopted major business fads such as total quality management (TOQ), velocity management (VM) in logistics and "just-in-time" logistics. These changes coincided with the increased influence of the information warfare enthusiasts who contended that that practices of so-called "new economy" can be applied to waging a war.[2]

Some prominent and highly influential U.S. military officials were apparently influenced by the book *War and Anti-War. Survival: At the Dawn of the Twenty-first century* written by Alvin and Heidi Toffler and published in 1993. The authors laid out several themes that were later accepted by the leading proponents of so-called network-centric warfare (NCW). The central theme of the Tofflers' work was that "the way we make war reflects the way we make wealth; and the way we make anti-war must reflect the way we make war."[3] They claimed that a revolutionary

---

1   Frederick W. Kagan, "A Dangerous Transformation. Donald Rumsfeld means business. That's a problem," *The Wall Street Journal*, November 12, 2003, p. 2, http://www.opinionjournal.com/extra/?id=110004289/

2   Ibid., p. 2.

3   Alvin and Heidi Toffler, *War and Anti-War: Survival At the Dawn of the 21st Century* (Boston, MA: Little, Brown and Company, 1993), p. 2.

"new economy" was arising based on knowledge, rather than conventional raw materials and physical labor. Supposedly, this remarkable change in the world economy is bringing with a parallel revolution in the nature of warfare.[4] Yet the nature of war as explained by Carl von Clausewitz is not subject to changes regardless of the changes in military technology not to say world's economy. This was one of the major errors in the Tofflers' book. The Tofflers also asserted that in "new economy" the time became a critical variable as reflected in "just-in-time" delivery and a pressure to reduce "decisions in process" (DIP). They were highly critical of those who opposed to the overreliance on technology in the U.S. military. The Tofflers expressed a clearly technological bias by arguing in favor of smaller numbers of highly sophisticated weapons and as exemplified the U.S./Coalition victory against Iraq in the Gulf War of 1990-1991. They wrote that in the new economies, the pace of operations and transactions is accelerated. The economies of speed are replacing economies of scale. Competition is so high and the speeds required so rapid that the old "time is money" rule is increasingly updated to "every interval of time is worthy more than the one before it."[5] The Tofflers also introduced the concept of "demassification" by asserting that the defining characteristics of the Second Wave economy become increasingly obsolete, as firms install information intensive, often robotized manufacturing systems capable of endless, cheap variation, even customization. The revolutionary result is, in effect, the demassification of mass production.[6]

By the late 1990s, the leading proponents of the then emerging network centric warfare concept embraced the Tofflers' idea that power flows from society and its methods of creating power and wealth. Hence, in their view the U.S. military should not read Clausewitz and other classical military thinkers but how the nations create wealth and prosperity.[7]

A major effort to extensively adopt various business models in the U.S. military was undertaken by Donald Rumsfeld during his tenure as the Secretary of Defense (2001-2006). His aim was to radically streamline the Pentagon by applying to the maximum various business practices. The logical outcome of Rumsfeld's approach would have been almost complete "homogenization" of all services. Each service would essentially have similar capabilities. This would lead to redundancies in their capabilities, which then, in turn, would be used as a justification for canceling additional weapon systems. The end result of this single-minded quest for military efficiency would be a much smaller but supposedly more mobile and lethal U.S. military force. The Pentagon also became enamored

---

4   Ibid., p. 5.
5   Ibid., pp. 63, 65.
6   Ibid., p. 59.
7   T.X. Hammes, "War isn't a Rational Business," *Proceedings*, July 1998, p. 23.

of outsourcing and just-in-time logistics, which eliminated supply depots and warehouses for spare parts.[8]

Network-centric warfare became the very heart of Rumsfeld's "Force Transformation" of the U.S. military. The leading advocates of Force Transformation repeatedly asserted that information revolution had fundamentally altered the ways of both business practices and the conduct of war. They explained that in business, success increasingly relied upon the ability to move material objects around. Business could produce items more rapidly and ship them faster and more cheaply. Those who did so were more successful than those that could not. Likewise, armies succeed by moving their forces to the decisive place and the time to defeat a similarly concentrated enemy army.[9] Business that could rapidly acquire, disseminate and analyze information would be more successful than the others. Network-centric warfare proponents argued that a fundamental shift in the sources of power—from industry to information—has already occurred, and that it is comparable to the earlier shift from the agrarian to the industrial age. In the information age, though industrial power remains important, information has become the most important source of power.[10] Yet the truth is that so-called "new economy" has not turned the law of supply and demand on its head. It did not represent more than the special features characterizing one of the periods of fundamental innovation that routinely occur in the economy.[11]

For network-centric warfare, leading proponents the Wal-Mart Corporation business practices were a model to be emulated by the U.S. military. For them, the main reason for Wal-Mart's success was its highly integrated system for gathering information at the point of sale allowed disseminating information not only to its own executives and other stores, but also to its suppliers.[12] In their view, modern military faced similar challenges. No longer would it be necessary to concentrate one's forces to achieve victory. Instead, success would belong to a side which acquires so-called "information dominance" and then conduct precise and lethal attacks by widely-dispersed platforms.[13]

---

8   Clay Risen, "The Danger of Generals-As-CEOs War-Mart," *The New Republic*, 28 (March 2006), p. 9.

9   Kagan, "A Dangerous Transformation. Donald Rumsfeld means business. That's a problem," p. 2.

10   Office of Force Transformation, *The Implementation of Network-Centric Warfare* (Washington, DC: Department of Defense, December 2005), p. 15.

11   Alfred Kaufman, *Curbing Innovation: How Command Technology Limits Network Centric Warfare* (Raleigh, NC: Scitech, 2004), p. 59.

12   Kagan, "A Dangerous Transformation. Donald Rumsfeld means business. That's a problem," p. 2.

13   Ibid., p. 3.

Purposes

The single most important difference between the conduct of war and business activity is their ultimate purpose and the ways of accomplishing these purposes. For one thing, the main purpose of any business is to create a customer and make profit.[14] In general, business activity follows certain rules and regulations. It has to conform to the existing social and legal order. In contrast, the ultimate purpose of warfare is not to create but, and this cannot be too strongly emphasized, to destroy the enemy's wealth and seize his territory and protect and preserve one's own. In contrast to business, war is full of violence and bloodshed. As Clausewitz aptly stated, war is an act of force, and the emotions cannot fail to be involved.[15] Whatever rules exist for its conduct are often violated by all sides. A wrong decision in business does not usually result in a loss of life. In contrast, a bad decision in war, especially one made by the top political leadership, is likely to result in huge losses in human lives and destruction of property. It might even have catastrophic consequences such as losing control of the territory or even foreign occupation and ultimately nation's very existence. Warfare is not simply making profits or avoiding losses and it is not primarily about preventing the waste of one's resources. A war involves the nation's vital interests—such as the nation's very survival and future well-being. War has to be won as quickly as possible, regardless of the costs involved.[16]

Art vs. Science

Common features of both the conduct of war and business management is each of them is both an art and a science. Business is an art because it requires creativity, intuition, and flexibility to motivate other people. Successful managers are able to develop unique alternatives and novel ideas about their organization needs. At the same time, business management is also a science. To be successful a good manager must draw on all the knowledge and insights of the humanities and the social sciences, on psychology and philosophy, on economics and history, on ethics as well as on the physical sciences.[17]

---

14   Peter F. Drucker with Joseph A. Maciariello, *Management*, revised edition (New York, NY: HarperCollins, 2008), pp. 97-8; Alfred Kaufman, *Curbing Innovation: How Command Technology Limits Network Centric Warfare* (Raleigh, NC: Scitech, 2004), p. 58.

15   Carl von Clausewitz, *On War*, translated by Michael Howard and Peter Paret (Princeton, NJ: Princeton University Press, 1976), p. 76.

16   Michael I. Handel, *Masters of War: Classical Strategic Thought*, 3rd revised and expanded edition (London: Frank Cass, 2001), p. 138.

17   Drucker with Maciariello, *Management*, p. 25.

Warfare is far more an art than business is. Arguably, of all the arts, warfare is also the most difficult.[18] A well-conducted war is like a great symphony. No good symphony is conceivable without many rehearsals and without uniform principles on which the whole orchestra is governed. At the same time, even in an excellently performed symphony, the musicians make major or minor mistakes causing discords.[19] Combat situations are extremely diverse and complex. They can change often and suddenly. These changes can be rarely anticipated either in terms of the time or scope. Clausewitz wrote that the "art of war must always leave a margin for uncertainty in the greatest things and in the smallest. The greater the gap between uncertainty on the one hand and courage and self-confidence on the other hand, the greater the margin that can be left for accidents."[20] The outcome of any war cannot be predicted with certainty, because so many intangible elements come into play.[21] War cannot be fought according to a preconceived scheme or using highly questionable scientific methods. In war, friction, danger, confusion, fear, fatigue, and discomfort, combined with a hostile physical environment, reduce the effective performance of both men and machines. Friction is inherent in any war. The more complex the technology, the higher the likelihood that something will not work as designed or will break down. As in the past, ambiguity, miscalculation, incompetence, and, above all, chance dominates the conduct of war.[22]

Human Factor

In both business and warfare, the human factors have a central and critical role. In management, the aim is to make people capable of working as a team and thereby enhancing their strengths while minimizing their weaknesses. To be successful, every business enterprise requires commitment to common goals and shared values. Management is about human beings. Its task is to make people capable of joint performance, to make their strengths effective and their weakness irrelevant. Business management is deeply embedded into culture. Every business

---

18   Friedrich von Boetticher, *The Art of War. Principles of the German General Staff in the Light of Our Time*, May 1951, ZA/1 2019 P-100, Bundesarchiv-Militaerarchiv (BA-MA), Freiburg, i. Br., p. 5.

19   Ibid., p. 6.

20   Carl von Clausewitz, *On War*, edited and translated by Michael Howard and Peter Paret (New York, NY/London/Toronto: Everyman's Library, Alfred A. Knopf, 1993), p. 97.

21   Paul K. van Riper and Robert H. Scales, Jr., "Preparing for War in the 21st Century," *Parameters* (Autumn 1997), pp. 2, 5.

22   Paul K. van Riper, "Information Superiority," statement before the Procurement Subcommittee and Research and Development Subcommittee of the House National Security Committee (March 20, 1997), pp. 2, 4, 5, 9-10.

enterprise requires commitment to common goals and shared values; without that commitment there is no enterprise. There is only a mob.[23]

The human element is the single most critical element of any warfare. In contrast to a business organization, the humans in the military live and work in close proximity with each other. There is far less room for one's privacy than is in a civilian life. The success of a military force in combat is largely dependent on the small-unit cohesion. The higher the cohesion of the tactical units, the higher the cohesion of large forces and formations taking part in a campaign or major operation.[24] A commander cannot be successful without thorough understanding of the capabilities and limitations of the human nature. Materiel represents the means not the ends in warfare. Warfare is too complex and unpredictable an activity to be taken over by machines. Only the human brain is fully capable of reacting promptly and properly to the sudden and unanticipated changes in the situation and counter the enemy's actions and reactions.

Clausewitz wrote that victory does not consist only in the conquest of the battlefield, but in the destruction of the enemy's physical and moral fighting forces. He believed in close linkage between morale and will power.[25] Because all wars are conducted by humans, the actions and reactions of the actors are hard or impossible to predict. This is even more true when dealing with the enemy forces. War is a field of danger.[26] Clausewitz observed that danger is "a part of the friction of war and without accurate conceptions of danger one cannot understand war."[27] The human behavior when faced with acute danger and fear cannot be anticipated or measured in any meaningful way. It is largely unknowable.

Rationality vs. Irrationality

The aim of both business and the conduct of war is to make rational decisions and act or react rationally. Economic theory is based on the assumption that all actors are rational. Nevertheless, irrationality plays a major part in economic behavior. Among other things, the markets are dominated by bubbles, fads and frenzies. Very often, the financial institutions and market traders take risks which they do not fully understand. Market operators can miscalculate, can be overly confident, and can overreact to bad news. For example, many people prior to the

---

23   Drucker with Maciariello, *Management*, p. 23.

24   John J. Johns et al., *Cohesion in the US Military: Defense Management Study Group on Military Cohesion* (Fort Lesley J. McNair, Washington, D.C.: National University Press, 1984), p. 4.

25   Cited in Beatrice Heuser, *Reading Clausewitz* (London: Random House, 2002), pp. 81, 84.

26   Alfred Stenzel, *Kriegführung zur See: Lehre vom Seekriege* (Hannover and Leipzig: Mahnsche Buchhandlung, 1913), pp. 40-41.

27   Clausewitz, *On War* (1993), p. 133.

U.S. recession in the fall of 2008 took on too much mortgage debt, which, in turn, was a major cause of the housing collapse. When the housing market was hot, bankers assumed that their customers did not want their houses to go into foreclosure and that they would act accordingly. The first assumption was correct, but the second assumption was flatly wrong.[28]

The rationality of the economic model assumes that investors react to changes in economic events assuming that they are always fully aware of the long-term implications of these changes, or that they have a superhuman vision.[29] There are the situations where individuals engaged in the economic activity act rationally, but the market might behave irrationally. The rational behavior on the part of individual investors can lead to collectively irrational outcomes or so-called "bubbles." For example, the famous "South Sea Bubble" in 1720 was based on the promise of unbounded riches to be garnered from trade with Spain's colonies in Latin America. Many investors knew that South Seas Trade had been exaggerated and that many of the bubble companies that issued stocks in the London market were fraudulent. However, they seized on the chance to make some quick money anyway.[30] The rational irrationality of the U.S. Wall Street insiders, in the form of "beauty contest" behavior, plays an important role in creating and sustaining bubbles. The theory of information cascades provides another example of how deliberate and purposeful behavior on the part of individuals can lead to collectively irrational results.[31]

In business activity, the relationship between rational individual and an irrational group of individuals can be extremely complex. One assumption is mob psychology or sort of group thinking when virtually all of the participants in the market change their views at the same time and move as a "herd." Alternatively, different individuals change their views about market development at different stages as part of a continuing process. Most of them start acting rationally but then more of them lose contact with reality, gradually at first and then more quickly. Another view is that different groups of traders, investors and speculators succumb to the hysteria as asset prices increase.[32] Periodic bouts of so-called "irrational exuberance" (coined in 1996 by Alan Greenspan, the former Chairman of the U.S. Federal Reserve Board) are endemic to the financial system.[33] Stock investors

---

28   Marshall Goldsmith, "Human Nature: The X Factor in Economic Theory," *Business Week*, January 20, 2009, http://www.businessweek.com/managing/content/jan2009/ca20090120_4029, p. 2.

29   Charles P. Kindleberger and Robert Z. Aliber, *Manias, Panics, and Crashes: A History of Financial Crises* (Hoboken NJ; John Wiley & Sons, 2005), p. 38.

30   John Cassidy, *How Markets Fail: The Logic of Economic Calamities* (New York, NY: Farrar, Straus and Giroux, 2009), p. 180.

31   Ibid., pp. 186, 190.

32   Kindleberger and Aliber, *Manias, Panics, and Crashes: A History of Financial Crises*, pp. 41-2.

33   Roger Bootle, *The Trouble with Markets. Saving Capitalism from Itself* (London and Boston, MA: Nicholas Brealey Publishers, 2009), p. 210.

would cause the market "bubble" through their greed and frenzy when bull market exists. This irrationality will, in turn, lead stock investors to overlook deteriorating situation because of their single-minded pursuit of ever-higher returns. Eventually, the frenzy of greed turns into the panic and thereby driving investors to sell at any cost. This collapse in stock market prices can spread to the entire economy.

One can presume that actors in a war make rational and proper choices when confronted with competing alternatives, each of which has a cost and a payoff or benefit that are known or available to the actors.[34] A rational calculus is based on the notion that nations fight wars in pursuit of postwar objectives whose benefits exceed the cost of their attainment. Costs and benefits are weighed throughout the war effort; and once the expenditures of effort exceed the scale of the political objective, the objective must be renounced and peace will follow.[35] The rationality of decision-making presupposes that each side knows exactly what the objectives of the other side are and what these objectives are worth in terms of efforts and sacrifices. Each also has all the necessary information to evaluate the other side's intent to continue or cease fighting. Thus, one or the other side can precisely calculate the enemy's relative current and future strengths. Also, one or both sides can identify and compare the anticipated costs of all available options.[36]

Clausewitz wrote that war is not the action of a living force upon a lifeless mass but the collision of two living forces that interact.[37] The enemy has his own will and will not behave the way one wants him to. He can react unpredictably and even irrationally. The timing and scope of irrationality cannot be either predicted or measured. The irrational decisions on either side in a war can have significant consequences on both the course and outcome of a war. For example, it is difficult or even impossible to rationally explain continuation of the hostilities on western front after 1916 for two more years despite huge losses in personnel and financial exhaustion.[38] Likewise, one cannot rationally explain why Hitler continued the war after 1943. It is also hard to rationally explain interminable inter-clan fighting in Somalia, genocide in Rwanda in 1994, or the Serbian ethnic cleansing in Bosnia between 1992-1995 and Kosovo in 1999.

Risk-Taking

Both a business executive and a military leader must take often some high risks in making decisions. The higher the level of authority and responsibility the higher

---

34   Michael I. Handel, *War, Strategy and Intelligence* (London: Frank Cass, 1989), p. 471.

35   Clausewitz, *On War* (1976), p. 92.

36   Michael I. Handel, *War Termination: A Critical Survey* (Jerusalem: Hebrew University, 1978), p. 29.

37   Clausewitz, *On War* (1993), p. 86.

38   Hammes, "War isn't A Rational Business," p. 23.

the stakes in taking risks. Business theory acknowledges the importance of risk. The opportunity cost of capital depends on the risk of the project. Reward, that is, the profit one makes, is determined by the risks one is willing to take. By failing to understand business risk one can make his business very vulnerable to sudden collapse. However, in contrast to the conduct of warfare, business theory postulates that the individual risk does not necessarily matter. Rather what matters is the risk in shares of similar businesses on the stock market adjusted for a further risk weighting. Some large businesses grow by transferring their business risk onto other people as is the case used in a buy-out model.[39]

Despite all the advances in information technologies, a commander will rarely know all the elements of the situation. This is especially the case at the operational and strategic levels of war. And it is at these levels where the wars are won or lost. In the absence of positive knowledge of the situation, the commanders must make certain assumptions that might by partially or completely wrong. Then he or she has to make decisions by taking a calculated risk. Willingness to take prudent risks means making operational decisions in varying degrees of uncertainty. Such decisions are critical for success, especially when the operational commander's forces are weaker than those of the enemy. They are not gambles, but carefully made, calculated decisions.[40] In contrast to the conduct of business, decisions made by the military commander can cause huge losses in one's personnel and materiel. Another difference is that the commander cannot share with or delegate risks to subordinate commanders. He or she is solely responsible for making decisions pertaining to planning, preparation and execution of campaigns or major operations.

Wal-Mart and Network-centric Warfare

The Pentagon became enamored with Wal-Mart's approach to business for understandable reasons. Network-centric warfare proponents described Wal-Mart as a self-synchronized distributed network with real-time transactional awareness. The stores' cash registers automatically transmit sales data to Wal-Mart's suppliers. The inventory is managed through horizontal networks rather

---

39   A leveraged buyout (LBO) or highly-leveraged transaction (HLT) (or "bootstrap" transaction) occurs when a financial sponsor acquires a controlling interest in a company's equity and where a significant percentage of the purchase price is financed through leverage (borrowing). The assets of the acquired company are used as collateral for the borrowed capital, sometimes with assets of the acquiring company. The bonds or other paper issued for leveraged buyouts are commonly considered not to be investment grade because of the significant risks involved.

40   Headquarters, Department of the Army, FM 22-103, *Leadership and Command at Senior Levels* (Washington, D.C.: U.S. Government Printing Office, June 21, 1987), p. 33.

than through a traditional head-office hierarchy.[41] The Wal-Mart conglomerate was successful because it used vast computer systems to lower inventories, respond better to consumer demand and even predict where their prospective markets are headed. The Wal-Mart system comprises three grids: infrastructure grid, sensor grid, and transaction grid. Infrastructure grid or sensor grid generates competitive space awareness while transaction grid exploits high levels of awareness to increase competitiveness. Competitive space awareness is a key competitive advantage in retail sector.

The network-centric warfare concept is essentially based on the Wal-Mart business model. Only the names of the three grids have been changed, to reflect the use of weapons and sensors. Battle space is considered as the military equivalent to Wal-Mart's "intelligent sales point." As explained by its proponents, network-centric warfare concept consists of three vertically linked grids: the sensor, shooter, and information grids. All three are interconnected—actions flow from sensors through decision-makers to shooters.[42] Network-centric warfare proponents contend that an enemy's attack on a single network should not result in the incapacitation of the whole.[43]

The *sensor grid* is composed of air-, sea-, ground-, space-, and cyberspace-based sensors. It provides the joint force with a high degree of awareness of friendly forces, enemy forces, and the environment across the joint battle space.[44] The *information grid* consists of a network of networks encompassing numerous communications paths, computational nodes, operating systems, and information management applications, allowing network-centric computing and communications across the joint battle space. It is designed to provide the means to receive, process, transport, store, and protect information for the multiservice (joint) and multinational (combined) forces. A permanent physical grid would exist in all mediums. The *shooter* (or *engagement*) *grid* consists of geographically dispersed air-, ground-, and sea-based shooters capable of delivering more responsive, accurate, and lethal fires.[45]

Network-centric warfare advocates contend that geographically dispersed forces embedded within an information network linking sensors, shooters, and command and control nodes would provide enhanced speed of decision-

---

41   Tomgram, Mike Davis on a 21st century Assyria with laptops, February 26, 2003, p. 2, http://tomdispatch.org/post/440/mike_davis_on_a_21st_century_assyria_with_laptops/

42   Wayne P. Hughes, Jr., *Fleet Tactics and Coastal Combat*, 2nd edition (Annapolis, MD: Naval Institute Press, 2000), p. 285.

43   John D. Zimmermann, "Net-Centric is about Choices," *Proceedings*, 1 (January 2002), p. 40.

44   Fred P. Stein, *Observations on the Emergence of Network-Centric Warfare,* (Vienna, VA: Evidence Based Research Inc, 1998), p. 2, http://www.dtic.mil/jcs/j6/education/warfare.html/

45   Ibid., pp. 2-4, 6-7, http://www.dodcrp.org.steincw.htm/

making, and rapid self-synchronization of the force as a whole to meet its desired objectives. They insist that these widely distributed and self-synchronizing forces would generate mass effects when and where desired.[46]

Network-centric warfare proponents also borrowed another term from the Tofflers' book—"demassification." Supposedly, netting of one's forces would allow planners and operators to go away from the methods based on geographically contiguous massing of forces to methods based upon achieving effects. NCW advocates explained that the use of information would lead to achieving desired effects, limiting the need to mass physical forces within specific geographic locations.[47] Substituting information and effects for mass would supposedly reduce the need to concentrate one's forces within specific geographical locations. This, in turn, would increase the tempo and speed of movement throughout the battle space, complicating an opponent's targeting problems.[48]

Combat Power and Network-Centric Force

The network-centric warfare advocates assert that one of the great benefits of netting one's forces is the significant increase in the forces' combat power. They assert that platform-centric warfare generates only "combat power," while network-centric warfare generates "increased power."[49] They repeatedly say that in the information era, power comes from information, access, and speed, while in the industrial era it came from mass.[50] However, it is extremely hard to measure the gain in the combat power of one's netted forces. The traditional elements of combat potential, such as raw firepower and mobility, are easier to assess. The problem of estimating true combat power is complicated by the presence of so many intangible factors that elude any quantification. In a networked force, all the gain in combat power can be significantly reduced and even nullified by micromanagement and excessive centralization. Also, poorly educated forces and incompetent commanders and staffs invariably reduce one's combat power.

The network-centric warfare advocates use a business analogy, asserting that the increased combat power of netted forces is the result of the application of Metcalfe's Law. However, they have misapplied the true meaning of that law to pervasive military networking. Metcalfe's Law pertains to the goods and services necessary to participate in a network. In its original meaning, Metcalfe's

---

46   Hughes, Jr., *Fleet Tactics and Coastal Combat*, p. 285.

47   Office of Force Transformation, *The Implementation of Network-Centric Warfare*, p. 9.

48   Arthur K. Cebrowski, *Military Transformation Strategic Approach* (Washington, D.C.: Office of Force Transformation, December 2003), p. 32.

49   Named after Robert M. Metcalfe, the inventor of Ethernet.

50   Paula R. Kaufman, "Sensors Emerge as More Crucial Weapons Than Shooters," *IEEE Spectrum Online*, July 16, 2003.

Law states that the value of a network increases with the square of the number of network users.[51] The "power" or "payoff" of network-centric computing comes from information-intensive interactions among very large numbers of heterogeneous computational nodes on the network.[52] However, the network-centric warfare enthusiasts changed "value" or "utility" to "power"—a highly questionable procedure. The most serious error is replacing the term "computers" with the word "computing" (a noun). In short, it might be true that Metcalfe's Law can be applied to networked computers but not necessarily to network computing. The network-centric warfare proponents stretched Metcalfe's Law even further by asserting that the power of transactions carried on a network increases with the square of the number of users of the network.[53]

The robustness of a network seems to imply that there is no upper limit to the accelerating benefits of additional network nodes. Hence, the larger the number of network nodes, the better. Metcalfe's Law justifies the quest to place the maximum possible number of battlefield actors on one interconnected network and thereby enjoy the military equivalent of the new economy's purported increasing returns. However, Metcalfe's Law actually breaks down at some sufficiently large n, at best flattening out the curve and at worst turning it back downward. In short, Metcalfe's Law does not promise increasing returns. Ultimately, the rate at which the network's value increases actually starts to decline with additional users. One reason for that is that as the use of the network increases; it becomes congested, making it less responsive as a means of transmission. In addition, as subscriptions to and use of the network increase, difficulties in conducting a search for relevant information increase. Delays in transmission and responsiveness due to congestion act as a potential drag on a network's value. Limitations on the capability of both humans and machines to assimilate and effectively process information constrain the growth of meaningful interaction. Not all networks have the same value. Also, not all nodes in a given network are of equal value. This undermines the argument in favor of ubiquitous networking. The most valuable nodes on a network tend to be occupied first; under these conditions, the rate of growth in the utility of the network may diminish or even reverse.[54]

Metcalfe's Law does not describe the gains to be obtained from network-enabled military interactions. There are adverse effects to networks. The benefit

---

51   Darryn J. Reid and Ralph E. Giffin, *A Woven Web of Guesses, Canto One: Network Centric Warfare and the Myth of the New Economy* (Washington, D.C.: 8th International Command and Control Research & Technology Symposium, National Defense University, June 2003), p. 6.

52   Arthur Cebrowski and John Garstka, "Network-Centric Warfare: Its Origins and Future," *Proceedings* (January 1998), p. 35; cited in Reid and Giffin, *A Woven Web of Guesses, Canto One: Network Centric Warfare and the Myth of the New Economy*, p. 7.

53   Reid and Giffin, *A Woven Web of Guesses, Canto One: Network Centric Warfare and the Myth of the New Economy*, p. 8.

54   Ibid., pp. 15, 17.

of military networking has its upper limit; the network-centric warfare thesis implies substantial centralization of authority and control.[55] When an organization joins a network, it surrenders its freedom in an important way. Lock-in gives the owner of the network significant power over the network's junior members. The real threat to the junior partner is the future decisions that the owner of the network is in principle free to make without consultation, limited only by the purely voluntary choice to take the new or junior members into consideration.[56]

Some theoreticians insist that the business analogy, far from supporting the network-centric warfare thesis, actually provides a stern warning against it.[57] With the advent of the Internet, the business world got a whole new medium of exchange, one that, among other things, broke down the traditional boundaries of the firm, allowing real-time collaboration irrespective of geography and offering the prospect of more effective alternatives to the traditional business structure. One of the inescapable consequences of traditional economic competition is the phenomenon of diminishing returns. As successful companies open up new markets, they typically attract competition, with the result that margins ultimately diminish. By positing that the value of a network increases with the square of the number of users of that network, Metcalfe's Law held out the prospect of a reversal of this margin-sucking trend in economics, at least for the increasingly important information technology sector of the new economy.[58]

Large vs. Smaller Forces

The network-centric warfare advocates contend that in the future armed forces will be much lighter and smaller. The logic behind this is their belief that dramatic advances in the precision and lethality of smart weapons will enable a major part of one's combat power to be brought to the battlefield from great distances. In theory, this would require a smaller presence of organic weapons. If fewer organic weapons are required, then the ground forces themselves can be made smaller and more dispersed. They would be harder for enemy forces to find and target. Supposedly, because of their smallness, the forces could be brought onto a battlefield quickly, even faster than conventional light airborne units. Relatively small and rapidly deployable forces would be capable of accomplishing missions that would otherwise require a large massed force.

Network-centric warfare enthusiasts contend that the new information technologies offer the potential to shift from large forces fighting sequential battles,

---

55   Ibid., pp. 20-21.

56   Ibid., pp. 12-13.

57   Ibid., p. 2.

58   Cebrowski and Garstka, "Network-Centric Warfare: Its Origins and Future," p. 35; cited in Reid and Giffin, *A Woven Web of Guesses, Canto One: Network Centric Warfare and the Myth of the New Economy* p. 4.

that is, attrition warfare, to near-simultaneous precision attacks by smaller forces. However, this and similar assertions are simply false. When properly applied in the past, operational art invariably led to quick and decisive victories and did not result in attrition warfare. Network-centric warfare proponents contend that smaller and more agile forces would have a significantly lower risk of taking casualties.[59] However, the penalty for having a small force is that, while it can be quickly deployed into the theater, it might not have sufficient combat potential to accomplish the assigned tasks.

Network-centric warfare enthusiasts also believe that netting geographically widely dispersed forces, combined with shared situational awareness and speed of communications, would allow much smaller forces to defeat much larger enemy forces, and very quickly. However, the network-centric warfare advocates' claim that relatively smaller forces can cover geographical areas because of their quick and flexible deployability is not supported by facts, as the examples of Afghanistan and Iraq illustrate.[60]

One of the factors that make network-centric warfare attractive to politicians and decision-makers is the prospect of smaller forces and the overall reduction in the numerical size of one's forces. However, the actual reduction in numbers is more applicable to air forces and, to a lesser extent, naval forces. The situation with ground forces is fundamentally different. A large number of troops might not be necessary in defeating weak and poorly armed opponents, such as the Taliban regime in Afghanistan or the much larger but quite ineffective Iraqi army. However, not all dictatorial and authoritarian regimes are brittle and ready to collapse, as Saddam Hussein's regime was. Unlike the forces of the other two services, ground forces are much more "people-centric" than "platform-centric," to use the technocrats' terms. The environment for land combat is also much more diverse and far more demanding than any other medium. Technological advances are unlikely to eliminate the requirements for close combat on the ground.[61] This is especially the case in unconventional warfare. The current situation in Iraq shows that information and technology cannot be substitutes for troops on the ground. The present situation is at least partly a direct consequence of the lack of sufficient forces in the major combat phase of the war against Iraq in 2003 (Operation Iraqi Freedom—OIF). Normally, much larger forces and different force mixes are required in the post-hostilities phase than in the major combat phase. Counterinsurgency efforts are by their very nature protracted and require the integrated use of not only military but also political, economic, informational,

---

59   William K. Lescher, "Network-Centric: Is it Worth the Risk?" *Proceedings*, 7 (July 1999), p. 59.

60   Aldo Borge, *The Challenges and Limitations of "Network Centric Warfare": The Initial Views of an NCW Skeptic* (Barton: Australian Strategic Policy Institute, September 17, 2003), p. 6.

61   Ibid., p. 8.

and other sources of one's power. All this requires much time and effort, and, above all, troops on the ground.

Business Model and Military Logistics

One of the key transformational concepts in Rumsfeld's era at the Pentagon was so-called "just-in-time" logistics. The main purpose was to reduce inventory to a minimum, and use of demand precision and optimization to reduce uncertainty. The proponents clearly believed that logistics planning is outdated. They claimed that demand is the true control signal in the logistic system containing more information about local operational conditions than a classic aggregation of supply.

U.S. forces used the "just-in-time" logistics concept during the invasion of Iraq in 2003. However, they encountered numerous difficulties due to poor logistical planning and overreliance on information technology. Logistical problems during the major combat phase of OIF included stretched supply lines during the rapid advance to Baghdad. If the Iraqis had offered much stronger resistance than they did, it would have posed a great risk to the coalition forces. Priority was given to the supply of fuel, ammunition, and food causing delays in supply of some critical spare parts. There was also some concern during the major combat phase of the operation that advancing units would run out of water, food, and fuel. Logisticians were often unable to accurately and timely distribute many items from ports to tactical units. Logistics units had inadequate communications and could not track times once items were removed from their shipping containers.[62]

Just-in-time logistics was an attempt to apply commercial practices to lean-out the inventory and make the logistic system more efficient. However, it can work properly in ideal conditions on the battlefield but not in the face of a determined enemy's opposition. It is inherently inflexible, vulnerable to damage, and unable service dynamically generated, prioritized needs. As the case of war in Iraq in 2003 showed, just-in-time-logistics as also brittle and very risky in a dynamic combat environment.

The U.S. military also adopted the commercial enterprise resources planning (ERP) concept to its logistics. The result was so-called "sense-and-respond" logistics (S&RL). This system is grounded in network-centric warfare theory and joint expeditionary warfare practice. It also borrowed from the commercial S&R adaptive managerial framework originally developed by IBM Corporation. The U.S. military transformation interpreted this concept by developing three major transformation efforts: network-centric operations (NCO), sense-and-respond logistics and military culture change.[63] Reportedly, the IBM model offered agile,

---

62   Kim Burger, "Iraq Campaign Raises New Logistics Concerns," *Jane's Defence Weekly*, September 10, 2003, pp. 16-17.

63   IBM Consulting Service, *Transforming the Military through Sense and Respond. White Paper*, January 2005, p. 1.

scalable, adaptable, and interoperable response capabilities. It is based upon the premise that changes in the business, security and technology environments are so rapid that they have outstripped one's ability to foresee and plan for them. A successful response would come from rapidly sensing and adopting to change rather than relying on process designs, hierarchies of authority and industrial age command-and-control action plans designed for more predictable events.[64] The main assumption of S&RL is that demand is ultimately unpredictable and hence success depends on speed of pattern recognition and speed of response; the best supply chain is highly flexible rather than highly optimized; organization of business units and subunits into modular capabilities that negotiate with one another over commitments. S&RL is highly dependent on sophisticated information technologies (IT) supporting to enable data sharing, knowing earlier, commitment tracking and role configuration. However, employment of one's combat forces during a major operation or conventional campaign is relatively short. Thus, the argument that one cannot properly plan for operational logistic support and sustainment is false. Among other things, the commander and planners must properly synchronize operations and logistics, otherwise a campaign or major operations might fail. This, in turn, requires thorough and timely logistical planning.

Efficiency vs. Effectiveness

The uncritical acceptance of some business models for the conduct of warfare by the U.S. military led to an increasing emphasis on efficiency rather than on effectiveness. *Efficiency* is the ratio of the output to the input into any system. It deals with one's skillfulness in avoiding waste of time and effort. A business can improve its bottom line by focusing on the few things it does very well and abandoning markets in which it is performing poorly. By eliminating redundancies and focusing on the areas in which they can excel, companies can dramatically improve their competitive position in some markets, even at the cost, of sometimes abandoning others. In business terms *effectiveness* is related to the enterprise's objective rather than the technical quality of output. A common indicator of effectiveness is related to customer satisfaction rather than output. Therefore effectiveness measure of a business process can be indicated by the resource inputs needed to produce a level of an enterprise objective.

In a military context, the effectiveness pertains to one's ability to accomplish the assigned objective—the starting point and a single most important element of both planning and execution in the employment of one's combat forces. Yet Rumsfeld's vision of the U.S. military transformation was focused almost

---

64   Ibid., pp. 1-2.

exclusively on efficiency rather than effectiveness.[65] For example, the U.S. Navy made a series of decisions regarding its force structure based almost entirely on the requirements of military efficiency rather than military effectiveness. Among other things, most newly built ships were assigned ever growing numbers of missions in order to build fewer platforms and thereby reduce the overall ships' construction costs. The SSNs are tasked to conduct ASW, covert surveillance/reconnaissance, antisurface warfare (ASUW), offensive mining, strikes against land targets, and insertion of Special Forces teams.[66] The primary missions of the new littoral combat ship (LCS) are to counter the threat of small boats, diesel submarines, and mines in the littorals. Its secondary missions include intelligence, surveillance, and reconnaissance (ISR), maritime intercept operations (MIO), homeland defense, support of special operations force (SOF) teams, and logistic support for movement of personnel and supplies.[67] Other missions currently being considered include fire support, carrying marines, and providing medical/humanitarian assistance. No one should dispute the need to have the highest degree of military efficiency in managing a large military organization and for force planning. However, when a choice has to be made, the military effectiveness should never be sacrificed to efficiency.

Leadership vs. Management

In generic terms, leadership can be described as the art of direct and indirect influence and the skill of creating the conditions for sustained organizational success in achieving desired results.[68] In contrast to leadership, management deals with the allocation and control of resources, whether human, material, or financial, in order to attain the objectives of an organization. Good management skills require neither an overabundance nor a shortage of resources.[69] The task of

---

65 Kagan, "A Dangerous Transformation. Donald Rumsfeld Means Business. That's a Problem," p. 3.

66 Ronald O'Rourke, *Navy Attack Submarine Procurement: Background and Issues for Congress* (Washington, DC: the Library of Congress, Congressional Research Service, updated April 8, 2008), p. CRS-2. Mark L. Gorenflo and Michel T. Poirier, "The Case for More Submarines," mhtml:file//J:\ARTICLES-ESSAYS-09 OCT 2008\ARTICLES IN JOURNALS\SCURR., p. 4.

67 Brien Alkire, et al., *Littoral Combat Ships. Relating Performance to Mission Package Inventories, Homeports, and Installation Sites*, p. 5; Ronald O'Rourke, *Navy Littoral Combat Ship (LCS) Program: Background, Oversight Issues, and Options for Congress* (Washington, DC: Congressional Research Service, June 5, 2009), p. CRS-2.

68 Headquarters, Department of the Army, FM 22-103, *Leadership and Command at Senior Levels* (Washington, DC: U.S. Government Printing Office, June 21, 1987), p. 3.

69 Cited in Brian Howieson and Howard Kahn, "Leadership, Management and Command: The Officer's Trinity," in *Airpower Leadership: Theory and Practice*, edited by Peter W. Gray and Sebastian Cox (London: The Stationery Office, 2002), p. 23.

management is to make people capable of joint performance through common goals, common values, the right structure, and the training and development they need to perform and to respond to change.[70] The superiority in materiel was one reason the U.S. military traditionally emphasized management thinking and a business approach to solving military problems. Among other things, the prominence of managerial values and entrepreneurial ethics was the main reason for the inability of the U.S. Army's officers to perform well in Vietnam.[71] Leadership is one of the most critical yet most complex aspects of warfare. The higher the level of command, the more important leadership skills are. The quality and skill of one's leadership cannot be measured; it is essentially intangible. No weapon, no impersonal piece of machinery ever designed can replace the human element in warfare. In contrast, militaries that traditionally emphasized leadership and warfighting, as the German military did, proved to be much more effective as a fighting force. The Germans focused on leadership as one way of enhancing combat potential and compensating for inferiority in materiel. Today's U.S. military excessive emphasis on management skill cannot but considerably erode the quality of leadership at all levels but especially at the higher command levels.

The Use of Business Metrics

Since the end of World War II, the Pentagon has used various quantifiable measures based on mathematical and statistical methods in trying to evaluate the effectiveness of bombing or ground forces involved in low-intensity conflict. It was not surprising that both McNamara and Rumsfeld, because of their business backgrounds, tried to use extensively business metrics in running the Pentagon. McNamara and his "whiz kids" believed that computers would transform the management of business. They invented a world where all decisions could be made based on numbers. They found power and comfort in assigning values to what could be quantified and deliberately ignoring everything else. McNamara brought analytical discipline to the military. Yet he also went too far by trying to conduct war by using Ford Company business model (where if an investment would not bring immediate profits it was vetoed). The whiz kids did not look far enough ahead with their cost-cutting calculations. They did not anticipate that they would lose customers and their engineering innovation in the long run.

McNamara firmly believed that the numbers can express almost any human activity. The things you can count you should count, he said. He applied statistical strategy of attrition in Vietnam. In his memoirs McNamara wrote that, in his view, it was not enough to determine an objective and a plan to carry out; you

---

70   Drucker with Maciariello, *Management*, p. 18.

71   Richard A. Gabriel, *The Antagonists: A Comparative Combat Assessment of the Soviet and American Soldier* (Westport, CT: Greenwood Press, 1984), p. 83.

must monitor the plan to determine whether you are achieving the objective. If you discover that you are not, you either revise the plan or change the objectives. He believed that the way to measure progress in Vietnam was to measure the targets destroyed in the North, the traffic down the Ho Chi Minh Trail, the number of prisoners,, the weapons seized, body count, and so on. The body count was a measurement of the enemy's losses. The Pentagon applied that method because one of the U.S. military objectives in Vietnam was to reach a so-called "crossover point," at which Vietcong and North Vietnamese casualties would be greater than they could sustain. Loss of life is a deceptively appealing measure when one is fighting a war of attrition. McNamara tried to use body counts as a measurement to help figure out what the U.S. should be doing in Vietnam to win the war while putting U.S. troops at the least risk.[72] Although such metrics proved to be meaningless he said that he did not have any regrets in using such quantifying methods.

In Vietnam, the American commanders tended to overstate their body counts. They did not have the means to identify what constituted a confirmed enemy kill. The Vietnamese tried to conceal from the Americans the number of their fighters killed. Despite huge number of the Vietcong fighters killed they were still able to find recruits and maintain a large force in the field. McNamara admitted that the Pentagon did not have correct information and hence improper quantifiable measures were used in evaluating combat in Vietnam. He also acknowledged that many aspects of the situation in Vietnam defied quantification and that "our misjudgment of friend and foe alike reflected our profound ignorance of the history, culture, and politics."[73] In short, the ability of the whiz kids for analysis far exceeded their knowledge of the world to which it was applied.[74]

The systems analysis and the use of other quantification methods in measuring progress on the battlefield fell into disrepute after McNamara left office in 1968. In the post-Vietnam era, the body count became irrelevant. However, the U.S. military reverted to the use of body counts in fighting a determined insurgency in Iraq and in Afghanistan.[75] The U.S. command in Afghanistan started in the early 2009 to publicize every single enemy fighter killed in combat. Supposedly the U.S. military adopted body count to undermine Taliban propaganda and stiffen the resolve of the U.S. public. The U.S. commanders in Afghanistan often have great difficulty acquiring precise information on the losses of Taliban. Very often

---

72 Robert McNamara with Brian Vandemark, *In Retrospect: The Tragedy and Lessons of Vietnam* (New York, NY: Times Books, Random House, 1995), pp. 237-8.

73 Ibid., pp. 238-9.

74 John Kay, "Managers doomed to repeat mistakes in history," *Financial Times Online*, FT.com, July 14, 2009, p. 2.

75 Tom Engelhardt, "Pentagon tracking a whole series of metrics in Iraq," *The Financial Express* (October 9, 2007), p. 1; available at http://www.thefinancialexpress-bd.com/2007/10/09/13819.html/

the Taliban remove bodies of their dead.[76] In contrast, the International Security Assistance Force (ISAF) in Kabul believes that such measurements are highly counterproductive in Afghanistan and unpalatable politically at home countries. The U.S. commanders in Afghanistan have often great difficulty in acquiring precise information on the losses of Taliban. Firefights often take place at ranges less than 1,000 yards and end with air strikes. Very often Taliban remove their dead.[77] To win the hearts and minds of the Afghans, Taliban are inflating the number of civilians killed and understating the number of their fighters killed. In short, body count is unreliable metric and should not be used in measuring progress of a war, especially in such a complex environment, as is counter insurgency where control of population, not the number of the enemy fighters killed, is the key for ultimate victory.

The excessive emphasis on business practices by the Pentagon has led since the late 1990s to an extensive reliance on various "metrics"[78] in evaluating the progress toward accomplishing the objectives on the battlefield. These quantification methods in essence replaced the commander's judgment, initiative and independence of execution.[79] There are too many aspects of the military situation, especially at the operational and strategic levels, that simply cannot be counted or quantified in any meaningful sense. The use of metrics is highly subjective because the higher authority arbitrarily selects which aspects of the situation should be counted and evaluated. But even if the metrics are correctly determined, it is often difficult to evaluate hidden elements in the situation. For example, in OIF the U.S. military used metrics such as sectarian murders and incidents within a specific time frame; weapons caches found by coalition and Iraqi forces; total attacks by province; average weekly attacks; average daily casualties; total national and regional hotline tips; confidence in the Iraqi government to improve the situation; the number of police stations lost to insurgents Iraqi; colonels and lieutenant colonels killed by the insurgents each

---

76   Michael M. Phillips, "Army Deploys Old Tactics in PR War," *Wall Street Journal*, June 1, 2009, p. A1.

77   Ibid., p. A1.

78   In business terms, a "metrics" is defined as any type of measurement, used to gauge some quantifiable component of a company's performance, such as return-on-investment (ROI), employee and customer churn rates, revenues (EB/TDA). They are a part of the broad area of business intelligence which comprises a wide variety of applications and technologies for gathering, storing, analyzing and providing access to data to help enterprise users to make better business decisions. Systematic approaches such as the balanced scorecard methodology can be used to transform an organization's mission statement and business strategy into specific and quantifiable goals and to monitor the organization performance in terms of achieving these goals.

79   Dieter Stockfisch, "Im Spannungsfeld zwischen Technologiefortschritt und Führungsverständnis. Auftragstaktik," *Marineforum*, 12 (December 1996), p. 13.

month; and others.[80] The Pentagon tracked whole series of metrics; some of them are inputs and some of them are outputs; metrics tracks included the number of attacks by area; the type of attacks by area; the number of reports of intimidation, attempts at intimidation and assassination of government officials; the extent to which people are supplying intelligence to our people; they also counting bodies as well.[81]

Critics of applying metrics in war in Afghanistan have pointed out that too many measures of progress have little or no value, report meaningless nationwide data, quantify the unimportant or are more designed to spin immediate success than to win real victory over time. Reportedly, the true complexities, uncertainties, and risks involved in dealing with a host of ethnic, sectarian, tribal and regional problems are downplayed or ignored.[82] One of the most damaging aspects of U.S. intelligence and advisory reporting is the tendency to focus on orders of battle which at best show manning levels and sometimes major equipment; it says little about unit progress and activity; what counts is what units do, how active they are, and how well they are led in actual operations.[83] Supposedly, measuring the nature and intensity of fighting and counts of the level and type of attack are still useful if they cover a full range of attacks by type, are broken out at least down to the province level, and are tied to the level of enemy progress or defeat in controlling the countryside. Overt violence is always an uncertain measure of insurgent activity and success.[84] Mapping control of the populating area—control of the population centers and space—requires different metrics. These key metrics include who governs and provides security in given areas and, especially, who controls an areas at night.[85] In broad terms, in a war like this, every area where the government does not actually govern—provide services at best—is vulnerable and often should be counted as lost.[86] Air coverage, air activity, and actual useful help, completing showpiece and demonstrating projects and aid efforts do not win hearts and minds; they lose them by telling those who were excluded.[87]

---

80   *Measuring Stability and Security in Iraq*, June 2007. Report to Congress In accordance with the Department of Defense Appropriations Act 2007 (Section 9010, Public Law 109-2890, pp. 17, 20, 21, 23-6.

81   Tom Engelhardt, "Pentagon tracking a whole series of metrics in Iraq," *The Financial Express*, October 9, 2007, p. 1, http://www.thefinancialexpress-bd.com/2007/10/09/13819.html/

82   Anthony Cordesman, *The Uncertain 'Metrics' of Afghanistan (and Iraq)* (Washington, D.C.: Center for Strategic and International Studies, May 18, 2007), p. 3.

83   Ibid., p. 6.

84   Ibid., p. 3.

85   Ibid., p. 4.

86   Ibid., p. 4.

87   Ibid., p. 6.

Business Vocabulary

By adopting the view that business practices can be successfully applied in the conduct of war, the U.S. military adopted also numerous business terms. This was particularly apparent in the writing of the leading proponents of network-centric warfare. They used extensively such business terms such as "transaction strategy," "competitive edge," "competitive space," "leveraging," "human capital strategy," "stakeholders," "VCNO Corporate Board," Navy enterprise construct" and its "warfare enterprises," "fleet readiness enterprise," "empowered self-synchronization," and so on. (Ironically, U.S. business is going in the opposite direction, increasingly adopting purely military terms in referring to their competitors.) The word "enemy" is being replaced with "threat," "adversary," or "opponent." The term "lock in our success" was adopted from the business term "product lock-in." Likewise, the term "battle space awareness" was adapted from the business term "competitive space awareness." The use of business terms is not only wrong but also creates a dangerous perception that warfare is not a messy and bloody affair but rather a nonviolent clash of the opposing interests.

Conclusion

There are a number of similarities between business activity and the conduct of war. The human factor plays a central and critical role in both business and warfare. The emotions, uncertainty, chance, and pure luck are characteristics of both business and warfare. Successful business managers and military commanders must often take calculated but high risks. Rationality and irrationality pervade decision-making and reactions in both business and warfare. However, for all the similarities, there are some significant differences between business and warfare. Clearly, the single most important distinction between the two is in their respective purposes. Management is much more important in business, while leadership counts far more in the conduct of war. Military effectiveness is the key for success in war, while efficiency is the primary consideration in making profits in business activity. Yet in their zeal to adopt business model, military technocrats focused almost exclusively on efficiency rather than military effectiveness. A Wal-Mart business model cannot be almost literally applied to the conduct of war as network-centric warfare enthusiasts did. Likewise, just-in-time and sense-and-respond concepts works well for business but might not be suitable for logistical support and sustainment in combat. There is no similarity in the conditions of the market place and on the battlefield. The errors or inability to bring timely certain items on the market do not result in the lives lost or property destroyed. Similar deficiencies of friendly forces with fuel, ammunition, and water can doom the military effort and result in large losses in lives.

By adopting various business metrics the U.S. military paid little or no attention to intangible factors in the military situation. Such quantification methods are not often unsuccessful even in business because the managers do not properly evaluate intangible factors on the market place. Metrics might have some limited utility in assessing the situation on the battlefield, but ultimately the success will be achieved by the decisions made by the commander based on his or her judgment and experience.

By its uncritical acceptance of some business models for the conduct of war, the U.S. military has neglected the critical and timeless importance of leadership and the human factor in the conduct of war. It also blurred and eroded the need for professional military judgment and eliminated the distinction between business activity and warfare. The U.S. military should use business practices whenever possible in enhancing the efficiency of the military establishment and services, force planning and designing weapons and equipment. However, using various business models in the planning and conduct of war itself and assessing the performance of one's forces in combat can have disastrous results as the U.S. experience in Vietnam shows. One can ignore lessons of history only at great peril.

Chapter 10

Cyberwar and Nuclear Crisis Management: Implications for Civil-Military Relations

Stephen J. Cimbala

Introduction

A favorable context for U.S. civil-military relations requires the deft management of both technology and policy development. The overlap between waxing and waning technologies relative to military art may require a delicate balancing act by policy makers. For example, the U.S. and other governments have continued to maintain and deploy nuclear weapons even as they pursue conventional force modernization based on an information-based Revolution in Military Affairs (RMA).[1] If the ultimate weapons of mass destruction—nuclear weapons—and the supreme weapons of soft power—information warfare—are commingled during a crisis, the product of the two may be an entirely unforeseen and unwelcome hybrid. Crises by definition are exceptional events. No Cold War crisis took place between states armed both with advanced information weapons and with nuclear weapons. But given the durability of the two trends, interest in infowar and in nuclear weapons, the potential for overlap and its implications for nuclear crisis management deserve further study and policy consideration. The discussion that follows considers pertinent concepts, policy related dilemmas and their implications for civil-military relations.

Concepts and Definitions

Cyber Omnipresent

The literature and the U.S. government already offer a rich menu of definitions for important cyber-related concepts, including cyberspace and

---

1 The concept of "Revolutions in Military Affairs" is examined with historical case studies in Colin S. Gray, *Strategy for Chaos: Revolutions in Military Affairs and the Evidence of History* (London: Frank Cass, 2002), especially ch. 8 on the nuclear RMA. See also: Max Boot, *War Made New: Technology, Warfare, and the Course of History, 1500 to Today* (New York: Gotham Books, 2006), pp. 307-436. Portions of the argument made in this chapter also appear in my article, "Nuclear Crisis Management and 'Cyberwar': Phishing for Trouble?," *Strategic Studies Quarterly*, 1 (Spring, 2011): 117-31.

cyberpower.[2] Information warfare can be defined as activities by a state or non-state actor to exploit the content or processing of information to its advantage in time of peace, crisis, or war, and to deny potential or actual foes the ability to exploit the same means against itself. This is an expansive, and permissive, definition, although it has an inescapable bias toward military- and security-related issues.[3] Information warfare can include both *cyberwar* and *netwar*. Cyberwar, according to John Arquilla and David Ronfeldt, is a comprehensive, information-based approach to battle, normally discussed in terms of high-intensity or mid-intensity conflicts.[4] Netwar is defined by the same authors as a comprehensive,

---

2   Pertinent U.S. government and other definitions for cyberspace and related concepts are reviewed in Daniel T. Kuehl, "From Cyberspace to Cyberpower: Defining the Problem," ch. 2 in *Cyberpower and National Security*, edited by Franklin D. Kramer, Stuart H. Starr, and Larry K. Wentz (Washington, D.C.: National Defense University Press – Potomac Books, Inc., 2009), pp. 24-42. See also, in the same volume: Martin C. Libicki, "Military Cyberpower," ch. 11, pp. 275-84, and Richard L. Kugler, "Deterrence of Cyber Attacks," ch. 13, pp. 309-40. Martin C. Libicki, *Cyberdeterrence and Cyberwar* (Santa Monica, CA: RAND Corporation, 2009), argues that strategic cyberwar is unlikely to be decisive, although operational cyberwar has an important niche role. Libicki also warns that deterrence in the cyber realm is unlikely to behave as it does in other domains, including conventional war and nuclear deterrence. See also: Will Goodman, "Cyber Deterrence: Tougher in Theory than in Practice?," *Strategic Studies Quarterly*, 3 (Fall, 2010): 102-35. Goodman argues that cyberspace poses unique challenges for deterrence but not necessarily impossible ones.

3   Concepts related to information warfare are discussed in David S. Alberts, John J. Garstka, Richard E. Hayes and David T. Signori, *Understanding Information Age Warfare* (Washington, D.C.: DOD Command and Control Research Program, U.S. Department of Defense, third edition October, 2004), especially pp. 53-94, and David S. Alberts, John J. Garstka and Frederick P. Stein, *Network Centric Warfare: Developing and Leveraging Information Superiority* (Washington, D.C.: Command and Control Research Program, U.S. Department of Defense, 6th printing April, 2005), especially pp. 87-122. Col. Thomas X. Hammes, USMC (Ret.), discusses the Pentagon's Joint Publication 3-13, *Information Operations*, and the U.S. Department of Defense understanding of information in modern warfare in Hammes, "Information Warfare," ch. 4 in *Ideas as Weapons: Influence and Perception in Modern Warfare*, edited by G.J. David, Jr. and T.R. McKeldin III (Washington, D.C.: Potomac Books, 2009), pp. 27-34. See also: John Arquilla, *Worst Enemy: The Reluctant Transformation of the American Military* (Chicago, IL: Ivan R. Dee, 2008), especially chs 6-7. For perspective on the role of information operations in Russian military policy, see Timothy L. Thomas, "Russian Information Warfare Theory: The Consequences of August 2008," ch. 4 in *The Russian Military Today and Tomorrow: Essays in Memory of Mary Fitzgerald*, edited by Stephen J. Blank and Richard Weitz (Carlisle, PA: Strategic Studies Institute, U.S. Army War College, July 2010), and Thomas, "Russia's Asymmetrical Approach to Information Warfare," ch. 5 in *The Russian Military Into the Twenty-first Century*, edited by Stephen J. Cimbala (London: Frank Cass, 2001), pp. 97-121.

4   Richard A. Clarke, former counterterrorism coordinator for the George W. Bush and Clinton administrations, and co-author Robert K. Knake include both cyberwar and netwar activities, as defined by Arquilla and David Ronfeldt, in their concept of "cyber

information-based approach to societal conflict. Cyberwar is more the province of states and conventional wars; netwar, more characteristic of non-state actors and unconventional wars.[5]

U.S. and Russian concepts of information warfare date from the Cold War years, although post-Cold War cyber, communications and electronics technologies have obviously required updating of operational concepts. China's determination to develop an informationized military was partly based on its assessment of U.S. success in Operation Desert Storm in 1991, which demonstrated superiority in technologies for C4ISR (command, control, communications, computers, intelligence, surveillance and reconnaissance), precision strike, and other attributes of advanced technology, information-based conventional warfare. Reportedly the People's Liberation Army (PLA) intends to develop a networked C4ISR system as part of its application of "network centric warfare" (NCW) to the Chinese military.[6]

Presumably, a network centric approach to military operations is a necessary part of any future PRC strategy for "anti-access, area denial" against U.S. or other forces intervening against PRC interests. On the other hand, these operational and tactical advances in information-based warfare do not necessarily summarize the entirety of China's needs or that of any other major power in this domain. Strategic information warfare at the highest level also includes the exploitation of information to deter or defeat attacks against the vital political and military centers of government and the armed forces, and, as well, to protect the critical parts of the civilian infrastructure without which the society and the economy cannot function.

---

war." See Richard A. Clarke and Robert K. Knake, *Cyber War* (New York: HarperCollins, 2010). For an introduction to this topic, see John Arquilla and David Ronfeldt, "A New Epoch—and Spectrum—of Conflict," in *In Athena's Camp: Preparing for Conflict in the Information Age*, edited by Arquilla and Ronfeldt (Santa Monica, CA: RAND, 1997), pp. 1-22. See also, on definitions and concepts of information warfare: Martin Libicki, *What Is Information Warfare?* (Washington: National Defense University, ACIS Paper 3, August 1995); Libicki, *Defending Cyberspace and Other Metaphors* (Washington: National Defense University, Directorate of Advanced Concepts, Technologies, and Information Strategies, February 1997); Arquilla and Ronfeldt, *Cyberwar is Coming!* (Santa Monica, CA: RAND, 1992); David S. Alberts, *The Unintended Consequences of Information Age Technologies: Avoiding the Pitfalls, Seizing the Initiative* (Washington, D.C.: National Defense University, Institute for National Strategic Studies, Center for Advanced Concepts and Technology, April 1996).

5   Arquilla and Ronfeldt, "The Advent of Netwar," in *In Athena's Camp*, pp. 275-94. With regard to the tasks for U.S. Cyber Command (established in 2009) and its implications for the national security decision making process, see Wesley R. Andrues, "What U.S. Cyber Command Must Do," *Joint Force Quarterly*, 59 (4th Quarter 2010), pp. 115-20.

6   Kevin Pollpeter, "Towards an Integrative C4ISR System: Informationization and Joint Operations in the People's Liberation Army," ch. 5 in *The PLA at Home and Abroad: Assessing the Operational Capabilities of China's Military*, edited by Roy Kamphausen, David Lai and Andrew Scobell (Carlisle, PA: Strategic Studies Institute, U.S. Army War College, June 2010), pp. 193-235.

According to Richard A. Clarke and Robert K. Knake, in this regard the U.S. is probably well ahead of peer competitors and other states in its ability to conduct offensive information warfare. However, the United States is also more dependent upon cyber systems and networks than some peer competitors or other prospective opponents are. Because of this higher network dependency factor and the difficulty of getting privately owned U.S. networks on the same page for cyber defenses, a "cyber war gap" exists that poses a potentially vital threat to U.S. security:

> When you think about "defense" capability and "lack of dependence" together, many nations score far better than the U.S. Their ability to survive a cyber war, with lower costs, compared to what would happen to the U.S., creates a "cyber war gap." They can use cyber war against us and do great damage, while at the same time they may be able to withstand a U.S. cyber war response. The existence of that "cyber war gap" may tempt some nation to attack the United States.[7]

In short, the race for strategic as well as operational-tactical military advantage in the information age is well under way. Information-based and information-enhanced sensors, shooter and commanders can turn OODA loops faster than their less info-empowered opponents, convert enemy computer networks into "botnets" of remotely controlled digital zombies, and implant surreptitious software behind enemy (digital) lines, awaiting future activation for the purpose of confusion or destruction. Under the best of circumstances, some of this might even be accomplished entirely in the cyber realm without kinetic overtures or leitmotifs: strategic or operational-tactical surrender, in the face of computer and network paralysis. Although things might turn out this way, the kinetic aspect of warfare, including the use of the most powerful weapons available for mass destruction, cannot be omitted as a consideration to be dealt with, amid the march of cyber tools and the relentless Revolution in Military Affairs (RMA) based on information and electronics.

In this regard, the very concept of "cyberdeterrence" involves degrees of uncertainty and complexity that require a leap of analytic faith beyond what we know, or think we know, about conventional or nuclear deterrence. Cyber attacks generally obscure the identity of the attackers, can be initiated from outside or within the defender's state territory, are frequently transmitted through third parties without their complicity or knowledge, and can sometimes be repeated almost indefinitely by skilled attackers, even against agile defenders. In addition, the contrast between the principles of cyberdeterrence and nuclear deterrence

---

7   Clarke and Knake, *Cyber War*, p. 149. The U.S. Defense Advanced Research Projects Agency (DARPA) is reportedly working on new cybersecurity programs potentially capable of learning during an attack and repairing themselves. See Cheryl Pellerin, "DARPA goal for cybersecurity: Change the game," American Forces Press Service, December 20, 2010, http://www.af.mil/news/story_print.asp?id=123235799/

encourages modesty in the transfer of principles from the latter to the former. As Martin Libicki summarizes:

> In the Cold War nuclear realm, attribution of attack was not a problem; the prospect of battle damage was clear; the 1,000th bomb could be as powerful as the first; counterforce was possible; there were no third parties to worry about; private firms were not expected to defend themselves; any hostile nuclear use crossed an acknowledged threshold; no higher levels of war existed; and both sides always had a lot to lose.[8]

Although experts might quibble about matters of degree, with respect to some of the preceding points, the case is clearly argued that, compared to nuclear deterrence, cyberdeterrence is a concept in search of further refinement by theoretical elaboration and empirical validation. For example, a division chief for the U.S. Army Global Network Operations Center laments "the lack of any meaningful cyberspace doctrine, or at least a serious consideration of how *cyberspace operations* differs from the closely related *computer network operations*, which is itself a key component of *information operations*."[9]

Airpower theorist and military analyst Benjamin S. Lambeth regards cyberspace as part of the third dimension of warfare that also includes air and space operations. Cyberspace, according to Lambeth, is the "principal domain" in which U.S. air services "exercise their command, control, communications, and ISR (intelligence, surveillance and reconnaissance) capabilities that enable global mobility and rapid long-range strike."[10] In addition, U.S. dominance, or falling behind, in cyberspace has repercussions for U.S. success or failure in aerospace and other domains of conflict.[11]

Added to this is the civil-military interaction that will take place between designated military cyber-samurai and their civilian DOD (and other) superiors in the chain of command who may be cyber-challenged or even pre-cyber in their understanding of information technology and its impacts. The nexus among new information capabilities, their implications for decision-making, and their potential vulnerabilities to attack may be comprehended by a select few, if at all.

Crisis Management

Crisis management, including nuclear crisis management, is both a competitive and cooperative endeavor between military adversaries. A crisis is, by definition,

---

8   Libicki, *Cyberdeterrence and Cyberwar*, p. xvi.

9   Andrues, "What U.S. Cyber Command Must Do," p. 115.

10   Benjamin S. Lambeth, "Airpower, Spacepower, and Cyberpower," *Joint Force Quarterly*, Issue 60, 1st quarter 2011, pp. 46-53, citation p. 50.

11   Ibid., p. 51.

a time of great tension and uncertainty.[12] Threats are in the air and time pressure on policymakers seems intense. Each side has objectives that it wants to attain and values that it deems important to protect. During a crisis state behaviors are especially interactive and interdependent with those of another state. It would not be too farfetched to refer to this interdependent stream of interstate crisis behaviors as a system, provided the term "system" is not understood as an entity completely separate from the state or individual behaviors that make it up. The system aspect implies reciprocal causation of the crisis behaviors of "A" by "B," and vice versa.

One aspect of crisis management is the deceptively simple question: what defines a crisis as such? When does the latent capacity of the international order for violence or hostile threat assessment cross over into the terrain of actual crisis behavior? A breakdown of general deterrence in the system raises threat perceptions among various actors, but it does not guarantee that any particular relationship will deteriorate into specific deterrent or compellent threats. Patrick Morgan's concept of "immediate" deterrence failure is useful in defining the onset of a crisis: specific sources of hostile intent have been identified by one state with reference to another, threats have been exchanged, and responses must now be decided upon.[13] The passage into a crisis is equivalent to the shift from Hobbes's world of omnipresent potential for violence to the actual movement of troops and exchanges of diplomatic demarches.

All crises are characterized to some extent by a high degree of threat, short time for decision, and a "fog of crisis" reminiscent of Clausewitz's "fog of war" that confuses crisis participants about what is happening. Before the discipline of crisis management was ever invented by modern scholarship, historians had captured the rush-to-judgment character of much crisis decision-making among great powers.[14] The influence of nuclear weapons on crisis decision-making is

---

12   For pertinent concepts, see: See Alexander L. George, "A Provisional Theory of Crisis Management," in *Avoiding War: Problems of Crisis Management*, edited by Alexander L. George (Boulder, CO: Westview Press, 1991), pp. 22-7, for the political and operational requirements of crisis management; and George, "Strategies for Crisis Management," ibid., pp. 377-94, for descriptions of offensive and defensive crisis management strategies. See also: Ole R. Holsti, "Crisis Decision Making," in *Behavior, Society and Nuclear War*, edited by Philip E. Tetlock et al. (New York: Oxford University Press, 1989), I, 8-84; and Phil Williams, *Crisis Management* (New York: John Wiley and Sons, 1976). See also Alexander L. George, "Coercive Diplomacy: Definition and Characteristics," in *The Limits of Coercive Diplomacy*, 2nd edition, edited by Alexander L. George and William E. Simons (Boulder, CO: Westview Press, 1994), especially pp. 8-9, and in the same volume, Alexander L. George, "The Cuban Missile Crisis: Peaceful Resolution Through Coercive Diplomacy," pp. 111-32.

13   See Patrick M. Morgan, *Deterrence: A Conceptual Analysis* (Beverly Hills, CA: Sage Publications, 1983); and Richard N. Lebow and Janice G. Stein, *We All Lost the Cold War* (Princeton, NJ: Princeton University Press, 1994), pp. 351-55.

14   For example, see Richard N. Lebow, *Between Peace and War: The Nature of International Crisis* (Baltimore: Johns Hopkins University Press, 1981); Michael Howard,

therefore not easy to measure or document because the avoidance of war can be ascribed to many causes. The presence of nuclear forces obviously influences the degree of destruction that can be done should crisis management fail. Short of that catastrophe, the greater interest of scholars is in how the presence of nuclear weapons might affect the decision-making process itself in a crisis. The problem is conceptually elusive: there are so many potentially important causal factors relevant to a decision with regard to war or peace. History is full of dependent variables in search of competing explanations.

Information Warfare and Nuclear Crisis Management: Possible Vulnerabilities

Information warfare has the potential to attack or to disrupt successful crisis management on each of the preceding attributes. First, information warfare can muddy the signals being sent from one side to the other in a crisis. This can be done deliberately or inadvertently. Suppose one side plants a virus or worm in the other's communications networks.[15] The virus or worm becomes activated during the crisis and destroys or alters information. The missing or altered information may make it more difficult for the cyber-victim to arrange a military attack. But destroyed or altered information may mislead either side into thinking that its signal has been correctly interpreted when it has not. Thus, side A may intend to signal "resolve" instead of "yield" to its opponent on a particular issue. Side B, misperceiving a "yield" message, may decide to continue its aggression, meeting unexpected resistance and causing a much more dangerous situation to develop.

Infowar can also destroy or disrupt communication channels necessary for successful crisis management. One way infowar can do this is to disrupt communication links between policymakers and military commanders during a period of high threat and severe time pressure. Two kinds of unanticipated problems, from the standpoint of civil-military relations, are possible under these conditions. First, political leaders may have pre-delegated limited authority for nuclear release or launch under restrictive conditions: only when these few conditions obtain, according to the protocols of pre-delegation, would military commanders be authorized to employ nuclear weapons distributed within their command. Clogged, destroyed, or disrupted communications could prevent top leaders from knowing that military commanders perceived a situation to be far more desperate, and thus

---

Studies in War and Peace (New York: Viking Press, 1971), pp. 99-109; Gerhard Ritter, *The Schlieffen Plan: Critique of a Myth* (London: Oswald Wolff, 1958); and D.C.B. Lieven, *Russia and the Origins of the First World War* (New York: St. Martin's Press, 1983).

15   A virus is a self-replicating program intended to destroy or alter the contents of other files stored on floppy disks or hard drives. Worms corrupt the integrity of software and information systems from the "inside out" in ways that create weaknesses exploitable by an enemy.

permissive of nuclear initiative, than it really was. For example, during the Cold War, disrupted communications between the U.S. National Command Authority and ballistic missile submarines, once the latter came under attack, could have resulted in a joint decision by submarine officers and crew to launch in the absence of contrary instructions.

Second, information warfare during a crisis will almost certainly increase the time pressure under which political leaders operate. It may do this literally, or it may affect the perceived time lines within which the policymaking process can make its decisions. Once either side sees parts of its command, control, and communications system being subverted by phony information or extraneous cyber-noise, its sense of panic at the possible loss of military options will be enormous. In the case of U.S. Cold War nuclear war plans, for example, disruption of even portions of the strategic command, control, and communications system could have prevented competent execution of parts of the SIOP (the strategic nuclear war plan). The SIOP depended upon finely orchestrated time-on-target estimates and precise damage expectancies against various classes of targets. Partially misinformed or disinformed networks and communications centers would have led to redundant attacks against the same target sets and, quite possibly, unplanned attacks on friendly military or civilian installations.

A third potentially disruptive effect of infowar on nuclear crisis management is that infowar may reduce the search for available alternatives to the few and desperate. Policymakers searching for escapes from crisis denouements need flexible options and creative problem-solving. Victims of information warfare may have a diminished ability to solve problems routinely, let alone creatively, once information networks are filled with flotsam and jetsam. Questions to operators will be poorly posed, and responses (if available at all) will be driven toward the least common denominator of previously programmed standard operating procedures. Retaliatory systems that depend on launch-on-warning instead of survival after riding out an attack are especially vulnerable to reduced time cycles and restricted alternatives:

> A well-designed warning system cannot save commanders from misjudging the situation under the constraints of time and information imposed by a posture of launch on warning. Such a posture truncates the decision process too early for iterative estimates to converge on reality. Rapid reaction is inherently unstable because it cuts short the learning time needed to match perception with reality.[16]

The propensity to search for the first available alternative that meets minimum satisfactory conditions of goal attainment is strong enough under normal conditions in nonmilitary bureaucratic organizations.[17] In civil-military command

---

16   Bruce G. Blair, *The Logic of Accidental Nuclear War* (Washington, D.C.: Brookings Institution, 1993), p. 252.

17   James G. March and Herbert A. Simon, *Organizations* (New York: John Wiley & Sons, 1958), pp. 140, 146.

and control systems under the stress of nuclear crisis decision-making, the first available alternative may quite literally be the last. Or, so policymakers and their military advisors may persuade themselves. Accordingly, the bias toward prompt and adequate solutions is strong. During the Cuban missile crisis, for example, a number of members of the presidential advisory group continued to propound an air strike and invasion of Cuba during the entire 13 days of crisis deliberation. Had less time been available for debate and had President Kennedy not deliberately structured the discussion in a way that forced alternatives to the surface, the air strike and invasion might well have been the chosen alternative.[18]

Fourth and finally on the issue of crisis management, infowar can cause flawed images of each side's intentions and capabilities to be conveyed to the other, with potentially disastrous results. Another example from the Cuban missile crisis demonstrates the possible side effects of simple misunderstanding and non-communication on U.S. crisis management. At the most tense period of the crisis, a U-2 reconnaissance aircraft got off course and strayed into Soviet airspace. U.S. and Soviet fighters scrambled, and a possible Arctic confrontation of air forces loomed. Khrushchev later told Kennedy that Soviet air defenses might have interpreted the U-2 flight as a prestrike reconnaissance mission or as a bomber, calling for a compensatory response by Moscow.[19] Fortunately Moscow chose to give the United States the benefit of the doubt in this instance and to permit U.S. fighters to escort the wayward U-2 back to Alaska. Why this scheduled U-2 mission was not scrubbed once the crisis began has never been fully revealed; the answer may be as simple as bureaucratic inertia compounded by non-communication down the chain of command by policymakers who failed to appreciate the risk of "normal" reconnaissance under these extraordinary conditions.

Other Implications

The outcome of a nuclear crisis management scenario influenced by information operations may not be a favorable one. Despite the best efforts of crisis participants, the dispute may degenerate into a nuclear first use or first strike by one side and retaliation by the other. In that situation, information operations by either, or both, sides might make it more difficult to limit the war and bring it to a conclusion before catastrophic destruction and loss of life had taken place. Although there are no such things as "small" nuclear wars, compared to conventional wars, there can be different kinds of "nuclear" wars, in terms of their proximate causes and

18   Richard N. Lebow and Janice G. Stein, *We All Lost the Cold War* (Princeton, NJ: Princeton University Press, 1994), pp. 335-6.

19   Graham T. Allison, *Essence of Decision: Explaining the Cuban Missile Crisis* (Boston, MA: Little, Brown, 1971), p. 141. See also Scott D. Sagan, *Moving Targets: Nuclear Strategy and National Security* (Princeton, NJ: Princeton University Press, 1989), p. 147; and Lebow and Stein, *We All Lost the Cold War*, p. 342.

consequences.[20] Possibilities include: a nuclear attack from an unknown source; an ambiguous case of possible, but not proved, nuclear first use; a nuclear "test" detonation intended to intimidate but with no immediate destruction; or, a conventional strike mistaken at least initially for a nuclear one.[21]

The dominant scenario of a general nuclear war between the United States and the Soviet Union preoccupied Cold War policy makers and, under that assumption, concerns about escalation control and war termination were swamped by apocalyptic visions of the end of days. The second nuclear age, roughly coinciding with the end of the Cold War and the demise of the Soviet Union, offers a more complicated menu of nuclear possibilities and responses.[22] Interest in the threat or use of nuclear weapons by rogue states, by aspiring regional hegemons or by terrorists, abetted by the possible spread of nuclear weapons among currently non-nuclear weapons states, stretches the ingenuity of military planners and fiction writers.

In addition to the world's worst characters engaged in nuclear threat or first use, there is also the possibility of backsliding in political conditions as between the United States and Russia, or Russia and China, or China and India (among current nuclear weapons states). The nuclear "establishment" or P-5 thus includes cases of current debellicism or pacification that depend upon the continuation of favorable political auguries in regional or global politics. Politically unthinkable conflicts of one decade have a way of evolving into the politically unavoidable wars of another—World War I is instructive in this regard. The war between Russia and Georgia in August, 2008 was a reminder that local conflicts on regional fault lines between blocs or major powers have the potential to expand into worse. So, too, were the Balkan wars of Yugoslav succession in the 1990s. In these cases, Russia's one-sided military advantage relative to Georgia in 2008, and NATO's military power relative to that of Bosnians of all stripes in 1995 and Serbia in 1999, contributed to war termination without further international escalation.

Escalation of a conventional war into nuclear first use remains possible where operational or tactical nuclear weapons have been deployed with national or

---

20   For pertinent scenarios, see George H. Quester, *Nuclear First Strike: Consequences of a Broken Taboo* (Baltimore, MD: Johns Hopkins University Press, 2006), pp. 24-52.

21   Ibid., p. 27.

22   Assessments of deterrence before and after the Cold War appear in: Patrick M. Morgan, *Deterrence Now* (Cambridge: Cambridge University Press, 2003); Colin S. Gray, *The Second Nuclear Age* (Boulder, CO: Lynne Rienner, 1999); Keith B. Payne, *Deterrence in the Second Nuclear Age* (Lexington, KY: University Press of Kentucky, 1996); Robert Jervis, *The Meaning of the Nuclear Revolution: Statecraft and the Prospect of Armageddon* (Ithaca, NY: Cornell University Press, 1989); and Lawrence Freedman, *The Evolution of Nuclear Strategy* (New York: St. Martin's Press, 1981 and 1983. Michael Krepon emphasizes that deterrence in the first nuclear age "worked," to the extent that it did so, only in conjunction with containment, diplomacy, military strength and arms control. See Krepon, *Better Safe than Sorry: The Ironies of Living with the Bomb* (Stanford, CA: Stanford University Press, 2009), passim.

coalition armed forces. In allied NATO territory, the U.S. deploys several hundred sub-strategic, air-delivered nuclear weapons among bases in Belgium, Germany, Italy, the Netherlands, and Turkey.[23] Russia probably retains several thousands of operational or tactical nuclear weapons, including significant numbers deployed in western Russia.[24] The New START agreement, once ratified, establishes a notional parity between the U.S. and Russia in nuclear systems of intercontinental range.[25] But U.S. and allied NATO superiority in advanced technology, information-based conventional military power leaves Russia heavily reliant on tactical nukes as compensation for comparative weakness in non-nuclear forces. NATO's capitals breathed a sigh of relief when Russia's officially approved Military Doctrine of 2010 did not seem to lower the bar for nuclear first use, compared to previous editions.[26]

Russia's military doctrine indicates a willingness to engage in nuclear first use in situations of extreme urgency for Russia, as defined by its political leadership.[27] And, despite evident superiority in conventional forces relative to those of Russia, neither the United States nor NATO is necessarily eager to get rid of their remaining sub-strategic nukes deployed among American NATO allies. An expert panel convened by NATO to set the stage for its 2010 review of the alliance's military doctrine was carefully ambivalent on the issue of the alliance's forward deployed nuclear weapons. The issue of negotiating away these weapons in return for parallel concessions from Russia was left open for further discussion. On the other hand, the NATO expert report underscored the present majority sentiment of

---

23   For detailed information on U.S. tactical nuclear weapons deployed in Europe, see Hans M. Kristensen, *U.S. Nuclear Weapons in Europe: A Review of Post-Cold War Policy, Force Levels, and War Planning* (Washington, D.C.: Natural Resources Defense Council, February 2005).

24   See Pavel Podvig, "What to do about tactical nuclear weapons," *Bulletin of the Atomic Scientists*, February 25, 2010, http://the bulletin.org, in *Johnson's Russia List* 2010, 43, March 3, 2010, davidjohnson@starpower.net, and Jacob W. Kipp, "Russia's Tactical Nuclear Weapons and Eurasian Security," *Jamestown Foundation Eurasia Defense Monitor*, March 5, 2010, in *Johnson's Russia List* 2010, 46, March 8, 2010, davidjohnson@ starpower.net, for pertinent insights and analysis.

25   *Treaty between the United States of America and the Russian Federation on Measures for the Further Reduction and Limitation of Strategic Offensive Arms* (Washington, D.C.: U.S. Department of State, April 8, 2010), http://www.state.gov/documents/organization/140035.pdf/

26   Text, "The Military Doctrine of the Russian Federation," www.Kremlin.ru/February 5, 2010, in *Johnson's Russia List* 2010, 35, February 19, 2010, davidjohnson@starpower.net. See also: Nikolai Sokov, "The New, 2010 Russian Military Doctrine: The Nuclear Angle," Center for Nonproliferation Studies, Monterey Institute of International Studies, February 5, 2010, http://cns.miis.edu/stories/100205_russian_nuclear_doctrine.htm/

27   See the analysis by Keir Giles, *The Military Doctrine of the Russian Federation 2010, NATO Research Review* (Rome: NATO Defense College, Research Division, February 2010), esp. pp. 1-2 and 5-6.

governments that these weapons provided a necessary link in the chain of alliance deterrence options.[28]

Imagine now the unfolding of a nuclear crisis or the taking of a decision for nuclear first use, under the conditions of both NATO and Russian campaigns employing strategic disinformation and information operations intended to disrupt opposed command-control, communications and warning systems. Disruptive information operations against enemy systems on the threshold of nuclear first use, or shortly thereafter, could increase the already substantial difficulty of bringing fighting to a halt before a Europe-wide theater conflict or a strategic nuclear war. All of the previously cited difficulties in crisis management under the shadow of nuclear deterrence pending a decision for first use would be compounded by additional uncertainty and friction after the nuclear threshold had been crossed.

Conclusion

Optimistic expectations about the use of information warfare to defeat or disrupt opponents on the conventional, high-technology battlefield—in cases where nuclear complications do not figure—may be justified. On the other hand, where the shadow of possible nuclear deterrence failure hangs over the decision-making process between or among states in conflict, the infowarriors' efforts to obtain dominant battlespace knowledge may provoke the opponent instead of deterring it. As scholars and policy analysts Keir A. Lieber and Daryl G. Press, have noted, with respect to U.S. superior performance at the sharp end of the conventional RMA: "A central strategic puzzle of modern war is that the tactics best suited to dominating the conventional battlefield are the same ones most likely to trigger nuclear escalation."[29]

The objective of infowar in conventional warfare is to deny enemy forces battlespace awareness and to obtain dominant awareness for oneself, as the United States largely was able to do in the Gulf War of 1991.[30] In a crisis with nuclear weapons available to the side against which infowar is used, crippling the foe's intelligence and command and control systems is an objective possibly at variance with controlling conflict and prevailing at an acceptable cost. And under some conditions of nuclear crisis management, crippling the C4ISR of the foe may be

---

28   *NATO   2020:   Assured   Security;   Dynamic   Engagement,   Analysis   and Recommendations of the Group of Experts on a New Strategic Concept for NATO* (Brussels: North Atlantic Treaty Organization, May 17, 2010), pp. 43-4.

29   Keir A. Lieber and Daryl G. Press, "The Nukes We Need: Preserving the American Deterrent," *Foreign Affairs*, 6 (November/December 2009): 39-51, citation, 43.

30   As David Alberts points out, "Information dominance would be of only academic interest, if we could not turn this information dominance into battlefield dominance." See Alberts, "The Future of Command and Control with DBK," in *Dominant Battlespace Knowledge*, edited by Stuart E. Johnson and Martin C. Libicki (Washington, D.C.: National Defense University, 1996), pp. 77-102, citation p. 80.

self-defeating. Deterrence, whether it is based on the credible threat of denial or retaliation, must be successfully communicated to—and believed by—the other side.[31] Whether nuclear or other deterrence can work in a particular context is more dependent upon political, as opposed to military, variables.[32]

As Mackubin Thomas Owens has emphasized, the essential problematique of U.S. civil-military relations is to combine the requirement for military subordination to civil authority with the need for military preparedness for conflict and effectiveness in combat.[33] The literature of civil-military relations, as in the case of other academic and policy studies, is mostly based in a pre-cyber world in which decision time and military operations moved more slowly than currently, and prospectively. In addition, cyber technology and information related concepts are becoming the critical enablers for everything else related to deterrence, war and preparations for war. Future warriors and political decision-makers will have to ensure that their military cyber experts are, as is sometimes asked of high powered lawyers, on tap, but never on top. Keep this point in mind as we move into a future of automated decision systems, artificial intelligence, UAVs, long range precision strike weapons, nanotechnologies, and improving capabilities for military exploitation of space.

---

31   As Colin S. Gray has noted, "Because deterrence flows from a relationship, it cannot reside in unilateral capabilities, behavior or intentions. Anyone who refers to *the* deterrent policy plainly does not understand the subject." Gray, *Explorations in Strategy* (Westport, CT: Greenwood Press, 1996), p. 33.

32   Lawrence Freedman, *The Evolution of Nuclear Strategy*, third edition (New York: Palgrave Macmillan, 2003), p. 463.

33   Mackubin Thomas Owens, *U.S. Civil-Military Relations after 9/11: Renegotiating the Civil-Military Bargain* (New York: Continuum, 2011), p. 12 and passim. The term civil-military *problematique* is attributed to Peter Feaver: Feaver, "The Civil-Military Problematique: Huntington, Janowitz, and the Question of Civilian Control," *Armed Forces and Society*, 23 (1966): 149-78.

Conclusion

Stephen J. Cimbala

The theory and practice of civil-military relations admit of considerable complexity. Plato's enduring question "Who shall guard the guardians?" is one that societies from the ancient to the modern have struggled to reconcile. Even within democratic political systems as a subset among all polities, the subordination of the military to civil power is a challenge to the integrity of the decision-making process, to the fundamentals of constitutional and common law, and to the armed forces' self-concepts of military professionalism and honor. As Mackubin Thomas Owens has noted:

> In many respects, the current state of theorizing about civil-military relations brings to mind the story of the three blind men examining an elephant. Since each can only sense what he is touching (the trunk, the leg and a tail), and has no concept of the elephant as a whole, each concludes that the beast is something different from what it really is. Despite the lack of an overarching framework for analyzing civil-military relations, the various areas of the field offer many rich "pastures" in which researchers many graze.[1]

Despite these complexities, soldiers and scholars have understandably felt the need for clarification and simplification of a larger reality in order to describe, explain and predict the character of civil-military decision-making processes and the outcomes of those processes. According to Peter D. Feaver, the conceptual pivot of civil-military relations or its *problematique* is the tension between two requirements. First, the armed forces of a state must be strong enough to prevail in war against other state or non-state actors that threaten the society those armed forces have been tasked to defend. Second, the military must not use its coercive power to obstruct civil control of the armed forces nor to prey upon the state and society that it protects.[2] As Feaver acknowledges, the tension between these two requirements exists in any political system or civilization, but it is "especially acute" in democracies. In democratic theory, the views of elected and appointed civilian policy makers must take precedence over those of their military establishment even when the views of the latter are, on technical and expert grounds, more right

---

1   Mackubin Thomas Owens, *U.S. Civil-Military Relations after 9/11: Renegotiating the Civil-Military Bargain* (New York: Continuum, 2011), p. 36.

2   Peter D. Feaver, *Armed Servants: Agency, Oversight, and Civil-Military Relations* (Cambridge, MA: Harvard University Press, 2003), pp. 4-5 and passim.

than wrong. As Feaver explains, democratic theory assigns different political and ethical competencies to civilian and military actors:

> The military officer is promising to risk his life, or to order his comrades to risk their lives, to execute any policy decisions. The civilian actor is promising to answer to the electorate for the consequences of any policy decisions. The military officer is expected to obey even stupid orders, or to resign in favor of someone who will. The civilian is claiming the right to be wrong. This forms a subtextual civil-military discourse for all policymaking in the national security realm.[3]

According to Feaver, Samuel Huntington's theory of U.S. civil-military relations remained as the dominant paradigm among contending approaches for many years because, among other reasons, it was firmly anchored in democratic theory.[4] Huntington's core arguments in favor of civil-military role differentiation, of military professionalism as the key to civilian control, and of military professional autonomy as the basis of professionalism, have been critiqued by other noted theorists.[5] Nevertheless Huntington's theory "survives despite repeated attempts to repeal it" while numerous challengers drift into obscurity.[6] Therefore, Feaver uses his critique of Huntington as a starting point for his own agency theory approach to U.S. civil-military relations.

A considerable literature notwithstanding, the dependent variable in the analysis of civilian control over the military is apt to be a reductionist one. One measure is whether civilian preferences prevail over military ones where there is obvious disagreement. Actual policy making in the U.S. system, and perhaps in other democracies as well, may make the distinction between "civilian" and "military"

---

3   Ibid., pp. 8-9.

4   Samuel P. Huntington, *The Soldier and the State: The Theory and Politics of Civil-Military Relations* (Cambridge, MA: Harvard University Press, 1957).

5   For a résumé, see Feaver, *Armed Servants*, pp. 7-9 and passim, and Owens, *U.S. Civil-Military Relations after 9/11*, especially pp. 12-43.

6   According to Feaver, those alternative approaches or critiques of Huntington (including sympathetic or congenial ones) that do endure are the exceptions that prove the rule. These include, for example: Morris Janowitz, *The Professional Soldier: A Social and Political Portrait* (New York: The Free Press, 1971, 2nd edition); Charles Moskos, John Allen Williams and David R. Segal, *The Postmodern Military: Armed Forces After the Cold War* (New York: Oxford University Press, 2000); Eliot Cohen, *Supreme Command: Soldiers, Statesmen, and Leadership in Wartime* (New York: Anchor Books, 2002); Michael Desch, *Civilian Control of the Military: The Changing Security Environment* (Baltimore, MD: Johns Hopkins University Press, 1999); Amos Perlmutter, *The Military and Politics in Modern Times: On Professionals, Praetorians, and Revolutionary Soldiers* (New Haven, CT: Yale University Press, 1977); Sam C. Sarkesian, *Beyond the Battlefield: The New Military Professionalism* (New York: Pergamon, 1981); and Sarkesian and Robert E. Connor, Jr., *The U.S. Military Profession into the Twenty-First Century: War, Peace and Politics* (London: Routledge, 2006). This list is obviously not exhaustive.

either superfluous or misleading. For many important U.S. national security policy decisions, one can find both military and civilian principals on the same side of an issue, opposed by another civilian-military coalition. Such was the case, for example, during the Cuban missile crisis of 1962, when hawks and doves within the presidentially empaneled "ExComm" for crisis deliberations were not necessarily identifiable by their civilian or military backgrounds and affiliation. Some civilian and military discussants favored an early ultimatum to the Soviets followed by an air strike and-or an invasion of Cuba. Other military and civilian ExComm members supported the "quarantine" or blockade option as a form of coercive diplomacy that would obtain U.S. objectives without military escalation. Some members of the ExComm changed their positions between the onset of the group's deliberations and their conclusion, and more than once.

During the Cold War and since, it has sometimes been the case that civilian leaders and their advisors were more eager for U.S. military intervention than were their military advisors or combatant commanders. Some research supports the idea that, in the U.S. case at least, military officers are more reluctant interveners in foreign conflicts than are their civilian counterparts—especially if the officers have had combat experience. Certainly General Colin Powell, then Chairman of the U.S. Joint Chiefs of Staff, was more reluctant than were civilian advisors of President George H.W. Bush to go to war against Iraq in 1991. Serving in the same position under President Bill Clinton, Powell was also hesitant about U.S. military intervention in Bosnia. In most cases, however, we would probably expect that military leaders would be more concerned with the "how" of military intervention than with the "whether" or "what" decision. Military planners and commanders want a clear mission statement, sufficient resources to accomplish their objectives, and some clear indications of Presidential and Congressional support for military intervention. The "Weinberger doctrine" that later evolved into the Powell doctrine was a matter of contention during the Reagan administration and has remained so into the present century. As a set of guidelines for when U.S. military intervention should, or should not, be undertaken, the "doctrine" is regarded by some as too restrictive (as against intervention in humanitarian disasters, for example). Others, including generations of military officers who experienced the Vietnam War and its postwar letdown during the sixties and seventies, favor the Weinberger principles as a litmus test for the political and military viability of a proposal for intervention.[7] In addition to the criterion of preference maximization across one or more national security issues, there is also the institutional aspect of recurring patterns in civilian

---

7   Pertinent appraisal of the Weinberger doctrine appears in Andrew J. Bacevich, "Elusive Bargain: The Pattern of U.S. Civil-Military Relations Since World War II, ch. 5 in *The Long War: A New History of U.S. National Security Policy Since World War II*, edited by Bacevich (New York: Columbia University Press, 2007), pp. 207-64, especially pp. 240-41, and in David H. Petraeus, "Lessons of History and Lessons of Vietnam," *Parameters* (Winter 2010-11), 40th Anniversary Issue: 48-61, especially 53 (originally published in Autumn 1986 issue of *Parameters*).

monitoring of military behavior and in military responses to that monitoring. Peter Feaver's agency model of U.S. civil-military relations, for example, emphasizes a two-dimensional matrix: (1) whether civilians monitor intrusively or non-intrusively; and (2) whether the military "works" or "shirks."[8] His model treats civil-military relations as a bargaining game of principal-agent relations. "Working" means that the military is doing things the way civilians want; "shirking" means that the military is acting to accomplish its own purposes regardless of civilian preferences.[9] The cross-classification of civilian monitoring styles with the military working-or-shirking dimension yields the following fourfold matrix:

Table 11.1   Agency Model and Traditional Theory

	Military works	Military shirks
Civilian monitors intrusively	Huntington's "crisis"	Extreme civil-military friction
Civilian does not monitor intrusively	Huntington's prescription	Lasswell's "garrison state"

Source: Peter D. Feaver, *Armed Servants: Agency, Oversight, and Civil-Military Relations* (Cambridge, MA: Harvard University Press, 2003), p. 120. Matrix simplified for use here.

Feaver admits that the terms working and shirking, when applied to the behavior of the U.S. armed forces, can appear as misleading and pejorative. Nevertheless they provide an economical dichotomy for assessing whether civilian or military preferences prevailed over time and in various issue areas. Based on this framework, Feaver's analysis of Cold War cases from 1945-1989 (decisions whether to use force and decisions how to use force) concludes that, for the most part, the military worked rather than shirked. Military preferences prevailed in only six of 29 cases for decisions "whether" to intervene, and in only four cases did the military thwart civilian preferences on "how" decisions. On balance, according to Feaver, U.S. civil-military relations during the Cold War belong in the "intrusive monitoring and working" cell and that the United States during the Cold War "enjoyed a remarkable degree of military obedience."[10] Similar conclusions, albeit from different theoretical perspectives, have been reached by Michael Desch, Lawrence Korb, and Allan Millett.[11]

---

8   Feaver, *Armed Servants*, pp. 58-64 and passim.

9   Ibid., p. 60.

10   Feaver, *Armed Servants*, p. 152.

11   Desch, *Civilian Control of the Military: The Changing Security Environment; Lawrence Korb, The Joint Chiefs of Staff: The First Twenty-Five Years* (Bloomington, IN: Indiana University Press, 1976), p. 131; and Allan R. Millett, *The American Political*

Richard K. Betts studied a number of Cold War incidents in which U.S. Presidents considered the commitment of armed forces to combat—from the Berlin blockade of 1948 to the "Christmas bombing" of Hanoi in 1972.[12] According to Betts, military advisors were usually divided in their recommendations on whether to commit military forces, and their recommendations followed civilian advice more often than they differed. As Betts explains:

> The diversity of military recommendations and the extent of consonance with civilian opinion indicate that military professionals rarely have dominated decisions on the use of force. Soldiers have exerted the greatest leverage on intervention decisions in those instances where they vetoed it. Disagreements with civilians were much greater on issues of the amount of force and the mode of implementation than on whether to commit American forces.[13]

Whether military advisors exercise excessive influence on military intervention or other decisions is of course only one aspect of the civil-military problematique. The other face of Janus is whether the military in a democratic society can perform its functional imperative (as Huntington called it) to provide for national defense and security. The first issue is one of military influence; the second, of military effectiveness.

According to Allan R. Millett, Williamson Murray, and Kenneth H. Watman, military effectiveness is "the process by which armed forces convert resources into fighting power."[14] The authors note that military effectiveness has both vertical and horizontal dimensions. The vertical dimension includes preparation for war and the conduct of war at the political, strategic, operational and tactical levels. The horizontal dimension refers to the "numerous, simultaneous, and interdependent tasks" that militaries must carry out at each hierarchical level in order to perform with proficiency.[15] The four levels of military activity (political, strategic, operational and tactical) are not separate compartments but overlapping categories. Nevertheless, each level has its own unique goals, actions and procedures, and military effectiveness must be assessed accordingly. Millett, Murray and Watman enumerate for each level various criteria for ascertaining the extent to which

---

System and Control of the Military: A Historical Perspective (Columbus, OH: Mershon Center, Ohio State University, 1979), p. 38.

12   Richard K. Betts, *Soldiers, Statesmen and Cold War Crises* (Cambridge, MA: Harvard University Press, 1977).

13   Ibid., p. 5. See also his summary tables in Appendix A, pp. 215-21.

14   Allan R. Millett, Williamson Murray, and Kenneth H. Watman, "The Effectiveness of Military Organizations," ch. 1 in *Military Effectiveness, Volume I: The First World War*, edited by Millett and Murray (Boston, MA: Unwin, Hyman, 1988), pp. 1-30, citation p. 2.

15   Ibid., p. 2.

armed forces have performed effectively in that dimension.[16] Significantly, these authors and Mackubin Thomas Owens emphasize that military effectiveness is not tantamount to "victory." Victory, according to Millett, Murray and Watman, is "an outcome of battle" and not "what a military organization does in battle."[17] As Owens puts it: military effectiveness may be a necessary requirement for victory, but not a sufficient cause for it.[18]

Definitions of victory or military effectiveness are unavoidably subjective no matter how many quantitative indicators are brought to bear on the analysis.[19] As Clausewitz and other great military thinkers have noted, war is relational: a contest of wills between two (or more) adversaries, each benchmarking its own subjectively driven concept of victory or defeat.[20] In addition, national cultures and societies influence their respective armed forces' and governments' definitions of victory and defeat, as well as honor and dishonor in the conduct of battle. Americans fighting in the Pacific theater in World War II learned, for example, that many Japanese officers and enlisted personnel were culturally and sociologically averse to the idea of surrender, even in the face of overwhelming odds and unavoidable catastrophe. In addition, many wars have shown that strategic or political surrender does not follow automatically from the defeat of the enemy's field forces. Wars can end from mutual exhaustion of the combatants or other causes for stalemate, with an armistice in lieu of a peace treaty that may follow later (World War I) or not at all (Korea). If an assassination attempt against Adolf Hitler had succeeded in July, 1944, and a post-Hitler German military regime had proposed a cease fire to the allied powers, would a negotiated peace with what remained of the Third Reich have made possible an earlier end to the war in Europe? Or, would the allies have remained steadfast in their commitment to "unconditional surrender" of the Axis regimes and their dismantlement, regardless the cost in blood and treasure? Arguably neither Franklin Roosevelt nor Winston Churchill would have treated diplomatically with any post-Hitler Reich: given prior allied sacrifice and the abominable character of Hitler's rule, only extirpation of the Nazi regime and any forces loyal to it would suffice for "victory."

Military or combat effectiveness is not necessarily any easier to define or to measure than the more abstract notion of "victory." As Sam C. Sarkesian has noted:

---

16   Ibid., pp. 4-30. See also the related discussion of military effectiveness by Owens, *U.S. Civil-Military Relations After 9/11*, ch. 3, pp. 90-93 and passim.

17   Millett, Murray and Watman, *Military Effectiveness Volume I*, p. 3.

18   Owens, *U.S. Civil-Military Relations After 9/11*, p. 92.

19   On definitions and concepts of victory, see William C. Martel, *Victory in War: Foundations of Modern Military Policy* (Cambridge: Cambridge University Press, 2007), especially pp. 94-103. His four-dimensional "pretheory" includes important distinctions among tactical, political-military and grand strategic victory.

20   Carl von Clausewitz, *On War*, edited and translated by Michael Howard and Peter Paret (Princeton, NJ: Princeton University Press, 1976), p. 75.

Achieving a high level of combat effectiveness is a function of the "art" of leadership, rather than a science. As most military historians will agree, success in battle and the performance of units in various combat situations are contingent on the will of the protagonists and the quality of leadership. Success usually means the imposing of the will of one side upon the other. Measurement must therefore parallel the "art" of achieving combat effectiveness.[21]

Definitions of military effectiveness and victory are also related to the kind of warfare that armed forces are tasked to fight. Some armed forces have missions that are limited to national self-defense and internal security. Other armed formations are not regular forces at all, that is, they are something other than a conventional military paid, provisioned and supervised by the government of a lawful state. These irregular armed formations are variously referred to as guerrillas, insurgents, terrorists or other labels designating them as apart from regular armies.[22] More than one twentieth-century conflict has been asymmetrical in the sense that at least one combatant force was regular and the other something else. It follows that unconventional and conventional conflicts have different aspects as to their political motives, combat operations, societal rationales, cultural implications, preferred command styles and societal costs. The U.S. armed forces, especially since the end of the Cold War and the demise of the Soviet Union, have aspired to full spectrum dominance in all aspects of warfare, but this aspirational standard has been tested in unconventional conflicts, including security and stability operations, counterinsurgency and counterterror missions, and nation building in the aftermath of regime change in Iraq and Afghanistan. As the expected spectrum of task competencies for U.S. forces has become progressively more elongated, a small but elite all-volunteer force has been strained to spread itself across a global community of trouble spots and international crises. An important U.S. Department of Defense Science Board task force concluded, with respect to ISR (intelligence, surveillance and reconnaissance) requirements for counterinsurgency:

While U.S. forces are superbly trained in their traditional aspects of violent combat, these irregular warfare and COIN campaigns may fail if waged by military means alone. Given that it is USG (U.S. Government) policy to deter and counter insurgencies, then the defense intelligence community should place more emphasis on "left of bang"—before the need for a large commitment of U.S. combat troops—while an insurgency is still in an incipient stage of

---

21   Sam C. Sarkesian, "Combat Effectiveness," Introduction, in *Combat Effectiveness: Cohesion, Stress, and the Volunteer Military*, edited by Sarkesian (Beverly Hills, CA: Sage Publications, 1980), p. 12.

22   Sorting these categories and their implications is accomplished in Richard H. Shultz, Jr. and Andrea J. Dew, Insurgents, *Terrorists and Militias: The Warriors of Contemporary Combat* (New York: Columbia University Press, 2006).

development, and that the whole-of-government approach should be given the capabilities necessary to succeed.[23]

Counterinsurgency and counterterror missions raise issues having to do with the interface among policy, intelligence and military operations that are not necessarily enlightened by the "civil-military" dichotomy. In the aftermath of 9-11, the covert actions of the Central Intelligence Agency became more paramilitary, as in the case of drone (UAV) attacks into Pakistan Taliban sanctuaries. At the same time, on the heels of experience in Afghanistan in 2001-2002, the Department of Defense established U.S. Special Operations Command as a "supported" command instead of a "supporting" command. USSOCOM could now plan and carry out its own missions with the approval of the Secretary of Defense and (when necessary) the President. USSOCOM was also designated in 2004 as the lead command in the war on terror, and, in the same year, Congress granted USSOCOM authority to spend funds for activities that previously required CIA approval (such as recruitment of foreign fighters and paying informants).[24] The U.S. military also increased its use of special operations forces as military liaison elements (MLEs), or small detachments of SOF located in embassies worldwide for gathering intelligence and disrupting terrorist operations. The successful cooperation between CIA and SOF in the early stages of Operation Enduring Freedom in Afghanistan in 2001 was an argument for synergy in improving both civilian intelligence and military covert action and special operations capabilities. On the other hand, as Gregory F. Treverton has noted, both SOCOM and CIA operations "raise thorny questions of authorization and accountability."[25] The operational synergy between civilian and military intelligence and special warfare activities was also apparent in President Obama's decision, announced in April, 2011 to nominate Gen. David Petraeus as the successor to CIA Director Leon Panetta, the latter designated to replace retiring Secretary of Defense Robert Gates. Petraeus, then U.S. and ISAF commander in Afghanistan, and formerly in charge of U.S. Central Command and the war in Iraq, had been proactive in mapping the intelligence requirements for military operations across the Middle Eastern and South Asian theaters, including special operations and counterterror missions, and was therefore an expert military "consumer" of intelligence.

Counterinsurgency is among the more ambitious kinds of military operations with which the U.S. has been faced since the end of the Cold War, but U.S.

---

23   Office of the Under Secretary of Defense for Acquisition, Technology and Logistics, *Report of the Defense Science Board Task Force on Defense Intelligence, Counterinsurgency (COIN) Intelligence, Surveillance, and Reconnaissance (ISR) Operations* (Washington, D.C.: U.S. Department of Defense, February, 2011), p. 24.

24   Gregory F. Treverton, *Intelligence for an Age of Terror* (Cambridge: Cambridge University Press, 2009), p. 229.

25   Ibid., p. 230.

experience with "COIN" in the twenty-first century is not new.[26] The United States was born in a revolutionary war and has since fought against irregulars that include Barbary pirates, native American tribes, marauders and brigands supporting the Confederate rebellion, Philippine insurrectos, Vietnamese nationalist guerrillas, and others. Periodically this experience and related "lessons learned" are resurrected and recycled into field manuals in the expectation that future generations of officers will not be forced to start anew on COIN or other unconventional missions. The current U.S. FM 3-24 captures the U.S. experience in Iraq and Afghanistan and provides doctrinal guidance for Army and Marine officers. However, it will disappear from their memory banks unless it is supported by hard experience, and that experience can only be acquired at the sharp end of a forward deployment in conflict zones penetrated by insurgents, terrorists and the like.[27] As then Major David H. Petraeus wrote with characteristic insight and uncommon prescience in 1986:

> Counterinsurgency operations, in particular, require close civil-military cooperation. Unfortunately, this requirement runs counter to the traditional military desire, reaffirmed in the lessons of Vietnam, to operate autonomously and resist political meddling and micromanagement in operational concerns. Military officers are of course intimately aware of Clausewitz's dictum that war is a continuation of politics by other means; many, however, do not appear to accept fully the implications of Clausewitzian logic.[28]

Regardless the military operational and tactical lessons to be learned from U.S. and allied coalition experience in Iraq and Afghanistan since 2001, those conclusions are part of a larger strategic and political context that is still in play. That context includes uncertainty about the extent to which the American and European public

---

26 See, for example, John A. Nagl, *Learning to Eat Soup with a Knife: Counterinsurgency Lessons from Malaya and Vietnam* (Chicago, IL: University of Chicago Press, 2005 edition), especially chs 2 and 8.

27 Expert assessments have grouped metrics for measuring progress in Afghanistan or other counterinsurgencies into four broad categories: population-related; host nation; local security forces, including army and police; and the enemy. See, for example: Ian S. Livingston, Heather L. Messera, and Michael O'Hanlon, *Afghanistan Index: Tracking Variables of Reconstruction and Security in Post-9/11 Afghanistan* (Washington, D.C.: Brookings Institution, May 31, 2011), especially p. 20, Figure 1.47, "Indicators for Measuring Progress in Afghanistan, Developed by David Kilcullen." See also: U.S. Defense Science Board Task Force on Defense Intelligence, Counterinsurgency (COIN) *Intelligence, Surveillance, and Reconnaissance (ISR) Operations*, pp. 75ff. Anthony H. Cordesman has produced a series of excellent reports on Afghan metrics and related narratives. See, for example, Cordesman, *Afghanistan and the Uncertain Metrics of Progress: Part Two: Transitioning to the New Strategy* (Washington, D.C.: Center for Strategic and International Studies, February 15, 2011).

28 Petraeus, "Lessons of History and Lessons of Vietnam," p. 54.

have steadfastness in the pursuit of counterinsurgency, counterterror, or peace and stability operations with possibly ambiguous goals and demanding of broad government competencies beyond the strictly military. This question is important because, like it or not, a few major powers with deployable field capabilities will determine when and whether military intervention is desirable and feasible. These powers may choose to act through multilateral or international organizations, such as NATO or the United Nations, but this does not change the equation of deployable military effectiveness for the prevention of humanitarian disasters, for regime change, or for other purposes. Some state or states must provide not only the military capacities demanded by the exigent situation, but also the additional competencies in social reconstruction, institution building, economic development, and, in all likelihood, inter-ethnic or inter-tribal conflict resolution. On the other hand, these additional competencies may be part of the problem, not the solution. Former Assistant Secretary of Defense and U.S. Marine combat veteran Francis J. (Bing) West critiques U.S. strategy in Afghanistan as too ambitious, with respect to governance, development and other objectives apart from security and combat. He recommends cutting back on the missions of population protection and democratic nation building in order to focus on neutralizing the enemy and on improved training for Afghan security forces.[29]

International intervention for the purpose of dampening or resolving internal wars in failing or failed states now takes place within the "information age" with all of its attendant baggage relevant to military affairs. The subtitles appearing in Lawrence Freedman's discussion of "strategic communications" speak volumes about the new world order of communications and information as it applies to conflict: the "information environment"; "media battles"; "hearts and minds"; "networks and hierarchies"; and so forth.[30] There is no longer a "nowhere" from the standpoint of media access should the leading networks and their reporters decide to provide coverage of a conflict. The Internet serves as a transmission belt for unofficial pictures and reportage crossing national borders and spanning the globe. Ready access to search engines, laptop computers and cell phones creates an info-sphere in which all, regardless of nationality or profession, are participant-observers in ongoing wars and national uprisings. Thus, for example, when street protests in 2011 began in Tunisia and intensified due to the global village audience provided by professional and impromptu media, authoritarian leaders across the Maghreb and elsewhere in the Middle East began to shake. Rulers in Tunisia and Egypt were unable to maintain any semblance of an "influence operation" and thus to persuade discontented middle class and educated protestors of the leaders' claim to rightful rule. Further shaking and baking of regimes by internal protest with

---

29   Bing West, *The Wrong War: Grit, Strategy, and the Way Out of Afghanistan* (New York: Random House, 2011), especially pp. 247-54.

30   See Lawrence Freedman, *The Transformation of Strategic Affairs*, Adelphi Papers 379 (London: International Institute for Strategic Studies, 2006), republished under the same title by Routledge (New York: 2006), ch. 5.

substantial media coverage soon appeared in Yemen, Bahrain, Syria and Libya. The Libyan case escalated into a civil war between anti-Qaddafi rebels and pro-regime factions that provoked international intervention by NATO, with United Nations approval.[31] Some of the skill sets required for commanders and their political leaders in these hybrid wars mixing conventional and unconventional military arts within a global info-sphere are noted by Freedman:

> Against this backdrop, governments and military commanders are bound to put considerable effort into describing, explaining and justifying operations. The ground might be prepared by feeding snippets of information, and operations may be timed with television schedules in mind. Spokespersons will need to avoid appearing unaware of blunders by their own side or risk being caught out in self-contradiction. A charismatic commander will be encouraged to help to convince a skeptical public as much as to boost the morale of front line troops. And so military operations have come to be understood in terms of the stories they tell as much as their direct impact on the enemy's physical capacity.[32]

Thus the dustup between the U.S. government and the info-insurgent group Wikileaks is not only about the potential for leaking classified material, but also about the stories that the material tells to foreign audiences, with respect to U.S. concepts of operations and war. Also in the case of photos from Abu Ghraib prison in Iraq that were leaked globally and produced a firestorm of recriminations within the U.S. government and media, they also "told a story" to some audiences in the Islamic community already predisposed to believe the worst about Operation Iraqi Freedom and coalition intentions. In another facet of the information war, terrorists and insurgents in Iraq and elsewhere have used portable media to record the results of their roadside bombings, ambushes and beheadings, circulating these snippets broadly for the purpose of favorably impressing politically committed and undecided observers.

Information wars occur in digitally accelerated time that precedes, accompanies and follows kinetic military activity. In the good old days of a few major news networks, governments could exercise some restraining influence over what was put out to the public at large by cajoling, wheedling, bribing or intimidating journalists and the owners of their enterprises. Not so today. The "mainstream media" know that if they cannot get a story out other media, including empowered individuals or interest groups, may "scoop" them and grab the initiative in

---

31   An insightful, if not skeptical, assessment appears in Anthony Cordesman, "Libya: Will the Farce Be With US (and France and Britain)?," April 20, 2011, burkechair@csis.org, also Center for Strategic and International Studies Website at http://csis.org/publication/libya-will-farce-stay-us-and-france-and-britain/

32   Ibid., p. 74. On the topic of influence operations, see John Arquilla, *Worst Enemy: The Reluctant Transformation of the American Military* (Chicago, IL: Ivan R. Dee, 2008), pp. 132-55.

storytelling. The traditional prestige media like the *New York Times*, past symbol of journalistic rectitude and unofficial public historian for the U.S. government, are driven by contemporary information technology and media competition into a niche market for policy wonks, political professionals, interest groups and other attentive publics. In the programming of what once were the major American "big three" news networks (ABC, CBS, NBC), survival now requires the commingling of what was formerly separated as "news" and "entertainment." In fact, so-called "news networks" have also mixed entertainment with news deliberately in order to raise ratings and therefore advertising revenues. Thus, "news" commentators on some of these networks are hired primarily for their appearances and chatterbox dexterity, regardless their command of the language or grasp of knowledge pertinent to news stories.

In addition to the amorphous, but powerful, impact of global media and information sphere on politics and warfare, there is the more specific new age menace of "cyberwar." The U.S. armed forces prefer the umbrella term "computer network operations" to include defense, attack and exploitation. These activities involve, respectively: protecting U.S. networks and systems from attack; attacking opposed systems in order to raise havoc by disrupting, destroying or degrading their information; and, third, exploiting information taken from opponents' networks and systems.[33] Cyberwar laps into civil-military relations because cyber attacks are anticipated against critical infrastructure, and most of that lies in the civilian realm. In addition, although cyber attacking and exploiting are related missions, in bureaucratic terms "attacking" is centered in the military, whereas "exploiting" is the province of the intelligence community (especially the U.S. National Security Agency).[34] Related constructs such as "information operations" or "influence operations" broaden the focus to include propaganda, disinformation, psychological operations and other approaches to manipulating the perceptions and expectations of opponents.[35] Even from the perspective of computer network operations per se, some experts have favored a broadening of focus to include

---

33   Robert E. Miller, Daniel T. Kuehl and Irving Lachow, "Cyber War: Issues in Attack and Defense," *Joint Force Quarterly*, 61, 2nd quarter 2011, pp. 18-23.

34   Ibid., p. 19.

35   Concepts related to information warfare are discussed in David S. Alberts, John J. Garstka, Richard E. Hayes and David T. Signori, *Understanding Information Age Warfare* (Washington, D.C.: DOD Command and Control Research Program, U.S. Department of Defense, third edition (October, 2004), especially pp. 53-94, and David S. Alberts, John J. Garstka and Frederick P. Stein, *Network-Centric Warfare: Developing and Leveraging Information Superiority* (Washington, D.C.: Command and Control Research Program, U.S. Department of Defense, 6th printing April, 2005), especially pp. 87-122. See also Col. Thomas X. Hammes, USMC (Ret.), "Information Warfare," ch. 4 in *Ideas as Weapons: Influence and Perception in Modern Warfare*, edited by G.J. David, Jr. and T.R. McKeldin III (eds) (Washington, D.C.: Potomac Books, 2009), pp. 27-34, and John Arquilla, *Worst Enemy: The Reluctant Transformation of the American Military* (Chicago, IL: Ivan R. Dee, 2008), especially chs 6-7.

"information and infrastructure operations" (I2Os) because opponents will attack both civil society and military systems.[36] The digitization of warfare will pose challenges for Clausewitzians with regard to the connection between force and policy, and for intellectual partisans of either purely objective or subjective civilian control over the armed forces. For example, cyberwars might very well: resist canonical notions of deterrence; begin with unclear objectives; defy unambiguous attribution of perpetrators; carry uncertain tendencies for escalation and escalation control; and, resist extant definitions of conflict termination and victory.[37]

The information or message war is not only one that needs to be fought "out there" in the military theater of operations. Popular support for U.S. war policy is the sine qua non upon which rests sustainable military power. It is therefore of some concern to students of civil-military relations that the superbly competent U.S. armed forces are in danger of becoming more disconnected from American society than hitherto. This is the sociological dimension of civil-military relations and it is related to, but distinct from, the military-operational or functional imperative of combat effectiveness. As Sam C. Sarkesian has explained, a society must provide for congruity between *military legitimacy* and *military posture*.[38] Military legitimacy has to do with the cohesion and purposes of the armed forces, as perceived by the society and by the military. Military posture has to do with the organization, training, technology and leadership of the armed forces. Military legitimacy, based mainly on subjective factors, is about the "rules of the game" that define the relationship between the armed forces and society. Military posture, based primarily on objective factors, is about combat proficiency and performance. According to Sarkesian, asymmetry between military legitimacy and military posture invites dysfunctional relationships between state and society and their armed forces. On one hand, if military legitimacy is too dominant over military posture, politicization of the armed forces and an erosion of military professionalism can result. On the other hand, if military posture is emphasized to the detriment of military legitimacy, the military perspective may come to dominate policy making—in the extreme situation, leading to a garrison state.[39] One need not resort to sociological abstraction to validate Sarkesian's case for harmony between military legitimacy and military posture. As Sebastian Junger has noted:

> War is a big and sprawling word that brings a lot of human suffering into the conversation, but combat is a different matter. Combat is the smaller game that young men fall in love with, and any solution to the human problem of war will

---

36 Miller, Kuehl and Lachow, "Cyber War," p. 19.

37 See Martin C. Libicki, *Cyberdeterrence and Cyberwar* (Santa Monica, CA: RAND Corporation, 2009), passim, and Miller, Kuehl and Lachow, "Cyber War," p. 21.

38 Sam C. Sarkesian, *Beyond the Battlefield: The New Military Professionalism* (New York: Pergamon Press, 1981), especially pp. 90-91.

39 Ibid., p. 91.

have to take into account the psyches of these young men. For some reason there is profound and mysterious gratification to the reciprocal agreement to protect another person with your life, and combat is virtually the only situation in which that happens regularly.[40]

According to some prominent military experts, the current and prospective relationship between the U.S. armed forces and American society is a strained one. Lawrence Korb, formerly assistant secretary of defense under President Ronald Reagan and senior scholar at the Center for American Progress, points to a "gulf" between military people and American society at a time when the country is engaged in two long wars.[41] According to Korb, there are three main reasons for this disconnect between the U.S. public and its armed forces. First, the U.S. has a total population of over 300 million people and a total military force (active and selected reserve) of 2.2 million. Even if U.S. force sizes were raised to Cold War levels of about 3 million, the armed forces would still constitute only about 1 per cent of the population. For comparison, during World War II American military forces counted 12 million in a nation with a total population of 132 million. Therefore, unlike the current situation, in the Second World War most Americans either knew someone in the armed forces or were related to a serving officer or enlistee.[42] A second reason for the public-military disconnect today, according to Korb, is that the connection was severed during the war in Vietnam because "many members of our political and financial elites managed to avoid being deployed to serve in Vietnam."[43] As a result, confidence in the Selective Service system vanished and the draft was replaced by an all-volunteer force. Third, according to Korb, after the 9-11 attacks, U.S. political and military leaders lost an opportunity to engage the public more directly on behalf of national policy by refusing to request publicly that the President and Congress activate the Selective Service System. In addition, neither the George W. Bush administration nor the U.S. Congress was willing to raise taxes to pay for the prolonged wars in Afghanistan and Iraq—to the contrary, taxes were cut as U.S. commitments deepened and casualties mounted. In sum, according to Korb:

> The wars in Iraq and Afghanistan are the first significant conflicts in our history for which we did not have a draft or raise taxes. Instead, we failed to activate the Selective Service system and actually cut taxes. Had we taken those steps, we would have moved in the direction of getting the country more involved and

---

40   Sebastian Junger, *War* (New York: Twelve – Hachette Book Group, 2010), p. 234.

41   Lawrence Korb, "The public's disconnect with the military," March 27, 2010, http://www.signonsandiego.com/news/2011/mar/27/the-publics-disconnect-with-the-military/

42   Ibid.

43   Ibid.

more aware of the sacrifices our troops are making even with a military that by nature of the size of the country would be a small part of our large population.[44]

One does not have to agree with Korb that pumping up the Selective Service system is a necessary part of a strategy for repairing the civil-military disconnection while, at the same time, sharing his concern about the civil-military divide in American society. In addition to the factors noted by Korb, other concerns include the disjunction between prevailing U.S. military notions of strategy and the popular (and Pentagon) preference for a small, elite and technologically based military. The American experience in fighting the "long war" against terrorism and related insurgencies in the past decade, in addition to the protracted wars in Afghanistan and Iraq, reminded us that protracted wars have a way of sucking up military deployments and exhausting the existing force, absent an expansive reserve base that can be tapped into in good time. Even the Reserve forces repeatedly deployed to Iraq and-or Afghanistan felt strained to the limits of their capabilities. Part of the problem was that, in the case of the U.S. Army in the latter stages of its involvement in Vietnam, force structure was redesigned so that critical mission capabilities were put into Reserve (including National Guard) units. Critics charge that U.S. Army Chief of Staff Gen. Creighton Abrams purposely did this in order to avoid any future Vietnams: any sizeable and sustainable commitment of forces to combat would require mobilization of reserve units, touching middle America and precluding the President from escalation by stealth from a small war into a larger. If Gen. Abrams did indeed plan his reorganization for this purpose, it shows a genius level grasp of American politics. It seems more likely that Abrams was trying to get more active-duty division equivalents in fighting power out of a force planning for downsized numbers in the early 1970s.

The preceding discussion shows that issues of strategy, structure, and political decision-making are connected within the civil-military "problematique," expanding it beyond a two-dimensional construct into a multidimensional one. The U.S. armed forces and civilian policy makers share a requirement and a responsibility to produce strategy at a high order of competency. Regardless the requirement, a strategy-friendly outcome or even a strategy-permissive one cannot be guaranteed to happen, under existing and foreseeable constraints imposed by American domestic politics. One has to be careful here of two dangers: exaggeration of the degree to which strategy departs from the optimal (the pessimist's temptation); and, the opposite fallacy, of celebrating U.S. abilities

---

44   Ibid. See also: Lawrence Korb and Laura Conley, "War in an age of deficits," CNNMoney, March 23, 2011, http://money.cnn.clom/2011/03/23'/news/economy/Libya_cost_lawrence_korb_/index.htm/

to "muddle through" despite the absence of anything resembling competent, or sometimes even tolerable, strategy making.[45]

Throughout the Cold War and subsequently, U.S. policy debates have emphasized decisions about weapons deployments, force sizes, budgets, operations and tactics, with comparatively less attention to the formulation of strategy. Strategic thinking is obviously part of the responsibility of the U.S. armed forces, but it is not theirs alone. Civilians with an appreciation for the unique character of the strategy "bridge" between policy and politics, on one side of the river, and military operations and tactics, on the other, are as necessary as are strategy-conscious warriors.[46] One has only to look at the exhaustive requirements for competent strategy making, including its political, economic, socio-cultural, technological, geographical and geostrategic, military, and historical dimensions, among others, to recognize that adequate, let alone superior, U.S. grand strategy cannot be accomplished by the military alone.[47] Civil-military relations are surely one aspect of national (grand) strategy, and equally important, national strategy cannot be expected to transcend the quality of a state's civil-military relations and its civil society—except under very exceptional conditions of grave threat or extraordinary leadership.

The preceding assertion suggests, for example, that Americans' "way of war" (and warfare) must take into account that definitions of military effectiveness and victory have a circumstantial provenance and an indeterminate half-life.[48] Strategic thinking however exemplary is subject to the smog effect created by 24/7 news networks, impatient publics, irresponsible policy makers, Internet cafe caudillos, and other distractions created by elite confusion and popular escapism. America as a culture is impatient for "results" that can be explained to masses in saccharine sound bites and to government officials in PowerPoint briefings. Thus we find that "shock and awe" is put forward as an ersatz descriptor for strategy, and "mission accomplished" is a video triumph despite considerable confusion about the mission itself. James L. Payne's historical and quantitative assessment of U.S. and British experience in "nation building" since 1850 offers the following judgment:

---

45   As Colin Gray has noted: "Strategy has just one function: to provide a secure connection between the worlds of purpose, which contestably is generally called policy, though politics may be more accurate, and its agents and instruments, including the military." Gray, *The Strategy Bridge: Theory for Practice* (Oxford: Oxford University Press, 2010), p. 29.

46   Ibid., p. 29 and passim.

47   See, Gray, *Modern Strategy* (Oxford: Oxford University Press, 1999), ch. 1 for an expansion on the topic of the dimensions of strategy. Gray lists 17 dimensions clustered into three categories (see especially p. 24).

48   Insightful commentary on this topic appears in James Kurth, "Variations on the American Way of War," ch. 2 in *The Long War*, edited by Bacevich, pp. 53-98.

Nation building by military force is not a coherent, defensible policy. It is based on no theory, it has no proven technique or methodology, and there are no experts who know how to do it. The record shows that it usually fails, and even where it appears to succeed, the positive result owes more to historical evolution and local political culture than anything nation builders might have done.[49]

Despite a cultural and political milieu unsupportive of strategy making, the U.S. armed forces have performed superbly well since the end of the Cold War, across a spectrum of conflicts from the unconventional to the coalition war against Iraq beginning in 2003. Perhaps the U.S. policy process is more permissive of military excellence than it appears, or perhaps the civil-military divergence that Korb (and others) have warned about is somehow healthy for growing armed forces that perform above their expectations. It would be interesting to see if the performances of other states' militaries are better than, or worse than, what we might expect from the strengths and weaknesses of their overall policy-making processes and the quality of their civil-military relations. Our contributors have put forward some interesting findings and arguments in this regard, some specific to individual countries, and others more generally applicable to states and armed forces. Regardless these insights and arguments, it remains the case that the relationship between armed force and civil power is both simple and complicated, in theory and in practice. Following Aristotle's distinction between the essence of a thing and its many possible attributes, civil-military relations can be manifest in infinite ways across time and among types of political societies. But the essence of civil-military relations in democratic societies lies in the ongoing renegotiation of the civil-military bargain as described by Mackubin Thomas Owens, in the "unequal dialogue" between civilians and military emphasized by Eliot Cohen, and the "constructive political engagement" advocated by Sam C. Sarkesian.[50] As Colin Gray has noted:

> If war is not to serve itself, with policy as its servant, there can be no argument
> over the primacy of the political over the military. Nonetheless, so close should
> be the dialogue between policy and its military agent that the strategist can
> hardly help but function to some variable degree both in a political or diplomatic
> role and to political effect. Similarly, the politician-policymaker, canonically a

---

49   James L. Payne, "Deconstructing Nation Building," *The American Conservative*, October 24, 2005, pp. 13-15, reprinted in Robert J. Art and Robert Jervis (eds), International Politics: Enduring Concepts and Contemporary Issues (Boston, MA: Longman, 2011), pp. 445-9, citation p. 449.

50   Owens, *Civil-Military Relations After 9/11*, pp. 12-13 and passim. Owens' bargain includes three parties: citizens, civilian government authorities and the military profession. See also: Cohen, *Supreme Command*, passim, and Sarkesian and Connor, *The U.S. Military Profession into the Twenty-First Century*, p. 74 and passim.

> civilian, certainly the authoritative representative of the civil power, cannot help but play as military strategist in his or her conduct of the "unequal dialogue."[51]

It is reasonable to ask whether the preceding arguments are based on case studies that over-represent democratic political systems compared to authoritarian ones. In addition, there are a variety of polities with both democratic and authoritarian aspects to their civil-military relations, as in Russia.[52] Armed forces in the twenty-first century will have to adjust their definitions of military legitimacy and military effectiveness to the exigencies of the information age—whether their political systems are authoritarian, democratic or mixed in nature. The requirements of the information age include new contexts for the application of military power as a means of political suasion. The operational-tactical and politico-strategic environments for the use or threat of force include land, sea, air, space and "cyber" or digital templates for warfare. New systems for long range precision strike, for C4ISTAR (command, control, communications, computers, intelligence, surveillance, targeting and reconnaissance) and stealth or low observables can be expected to challenge planners of conventional and unconventional wars alike. In addition, competency in the soft power aspects of the "war of ideas" (possibly including some aspects of cyber war) will demand cooperation among military arms of service within a state, collaboration among the armed forces of coalitions, as in Afghanistan and Iraq, and inter-agency cooperation within and among states—including the willingness of governments to work with NGOs and other entities that contribute to the legitimation and success of a mission.[53]

For high-end militaries such as those of NATO, Russia and China, even in the case of focused counterterror operations (presumably less ambitious than more protracted and costly counterinsurgencies), the necessity to mix hard and soft, high tech and low tech, and messages of deterrence with those of reassurance, will confound standard operating procedures in defense and foreign ministries. More scripts for operations and the policies to support them will be written as open-ended narratives with branches and fewer reduced to the simplicity of PowerPoint options. For example, the U.S. raid into Pakistan to kill Osama bin Laden in May, 2011 combined advanced technology (stealth helicopters!) and command-control systems

---

51   Gray, *The Strategy Bridge*, p. 204.

52   See, in addition to Stephen J. Blank's chapter in this volume, see: Lilia Shevtsova, *Russia – Lost in Transition: the Yeltsin and Putin Legacies* (Washington, D.C.: Carnegie Endowment for International Peace, 2007), especially pp. 47-65 on "imitation democracy;" and Anders Aslund and Andrew Kuchins, *The Russia Balance Sheet* (Washington, D.C.: Peterson Institute for International Economics and Center for Strategic and International Studies, 2007), pp. 30-38, 103-14, and 119-20.

53   As Colin Gray has noted: "There is strategic advantage in moral advantage, which translates as a requirement for the use of military force to be plainly legitimate." *Gray, Hard Power and Soft Power: The Utility of Military Force as an Instrument of Policy in the 21st Century* (Carlisle, PA.: Strategic Studies Institute, U.S. Army War College, April 2011), p. 50.

with gut-wrenching derring-do by Navy SEALS, including backup helicopters for rescue and the inevitable "friction" of combat. Information operations kept a very small circle of U.S. leaders and military planners "in the loop" in order to prevent leaks, including leaks to distrusted factions in Pakistan. On the other hand, following the successful conclusion of the military operation, some White House after-action reports were misleading or confusing until inter-agency coordination and additional scrutiny of available evidence clarified matters. The intense worldwide press interest in pertinent revelations and video probably led to some premature statements later retracted or modified.

If authoritarian governments and their militaries think they will escape the pressures of technology modernization and media blitzkrieg, they are mistaken. A globally networked media and digital commons will turn transparently absurd policy statements about national security into transnational amusements on Twitter and Facebook if not on the BBC and CNN. This does not mean "democracy everywhere" but it does mean that transparency, with progressively more relentlessness, stalks the security policies and civil-military relations of authoritarian polities, of democracies, and of hybrids. In a digital media world where small mistakes appear as strategic blunders, and strategic blunders as apocalyptic endings, governments and regimes will progress at accelerating rates from political gravitas to political graveyard. The victories of their armed forces will be quickly forgotten and their defeats greatly magnified. Especially in the case of unconventional conflicts, where combat becomes one instrument of influence among many, and not necessarily the most important, clear definitions of victory or defeat will be replaced by oblique statements of objective and accomplishment. Compared to the previous century, in the twenty-first century relatively fewer wars may be terminated decisively and rapidly, especially in their social and cultural aspects. Chechens have been fighting Russians in the Caucasus for more than a century and, even if Chechnya were suitably pacified by Russian standards, indigenous or imported jihadisms will continue to stalk southern Russia. Or, in the case of the United States and NATO in present-day Afghanistan, the expectation of clear and decisive military victory may give way to a more achievable political endgame, permitting large-scale NATO troop withdrawals and their replacement by Afghan national army and police forces. However the exit strategy is timed and carried out, we can anticipate that considerable controversy will mark domestic policy debates on both sides of the Atlantic. These debates will hold important implications for U.S. and allied civil-military relations. The kinds of wars we fight and the ways in which we choose to fight them, and end them, are the base ingredients for future definitions of military legitimacy and military effectiveness.

Selected Bibliography

Arquilla, J., 2008. *Worst Enemy: The Reluctant Transformation of the American Military* (Chicago, IL: Ivan R. Dee).

Åslund, A. and A. Kuchins, 2007. *The Russia Balance Sheet* (Washington, D.C.: Peterson Institute for International Economics and Center for Strategic and International Studies).

Bacevich, A.J., 2007. "Elusive Bargain: The Pattern of U.S. Civil-Military Relations Since World War II," ch. 5 in *The Long War: A New History of U.S. National Security Policy Since World War II*, edited by A.J. Bacevich (New York: Columbia University Press), pp. 207-64.

Bacevich, A.J., 2010. "Endless war, a recipe for four-star arrogance," *Washington Post*, June 27, p. B01, http://www.washingtonpost.com/wp-dyn/content/article/2010/06/25/AR2010062502160_pf/

Betts, R.K., 1977. *Soldiers, Statesmen and Cold War Crises* (Cambridge, MA: Harvard University Press).

Bevin, A., 2002. *How Wars Are Won: The 13 Rules of War: From Ancient Greece to the War on Terror* (New York: Crown Publishers).

Blank, S.J., 2010. "No Need to Threaten Us, We Are Frightened of Ourselves," Russia's Blueprint for a Police State, The New Security Strategy," ch. 1 in *The Russian Military Today and Tomorrow: Essays in Memory of Mary Fitzgerald*, edited by S.J. Blank and R. Weitz (Carlisle, PA: Strategic Studies Institute, U.S. Army War College, July) pp. 19-149.

Clausewitz, C. von, 1976. *On War*, edited and translated by M. Howard and P. Paret (Princeton, NJ: Princeton University Press).

Cohen, E.A., 2002. *Supreme Command: Soldiers, Statesmen, and Leadership in Wartime* (New York: Free Press).

Cordesman, A., 2011. "Libya: Will the Farce Be With US (and France and Britain)?," April 20, burkechair@csis.org, also Center for Strategic and International Studies Website at http://csis.org/publication/libya-will-farce-stay-us-and-france-and-britain/

Desch, M., 1999. *Civilian Control of the Military: The Changing Security Environment* (Baltimore, MD: Johns Hopkins University Press).

Feaver, P.D., 1992. *Guarding the Guardians: Civilian Control of Nuclear Weapons in the United States* (Ithaca, NY: Cornell University Press).

Feaver, P.D., 2003. *Armed Servants: Agency, Oversight, and Civil-Military Relations* (Cambridge, MA: Harvard University Press).

Feaver, P.D. and C. Gelpi, 2004. *Choosing Your Battles: American Civil-Military Relations and the Use of Force* (Princeton, NJ: Princeton University Press).

Flavin, W., 2011. *Finding the Balance: U.S. Military and Future Operations* (Carlisle, PA: Peacekeeping and Stability Operations Institute (PKSOI), U.S. Army War College), March.

Freedman, L., 2006. *The Transformation of Strategic Affairs*, Adelphi Papers 379 (London: International Institute for Strategic Studies —Routledge Publishers).

Godson, R. and R.H. Shultz, Jr., 2010. "A QDR for All Seasons? The Pentagon Is Not Preparing for the Most Likely Conflicts," *Joint Force Quarterly*, 59 (4th Quarter), pp. 52-6.

Gray, C.S., 1999. *Modern Strategy* (Oxford: Oxford University Press).

Gray, C.S., 2010. *The Strategy Bridge: Theory for Practice* (Oxford: Oxford University Press).

Gray, C.S., 2011. *Hard Power and Soft Power: The Utility of Military Force as an Instrument of Policy in the 21st Century* (Carlisle, PA: Strategic Studies Institute, U.S. Army War College), April.

Hanson, V.D., 2010. *The Father of Us All: War and History, Ancient and Modern* (New York: Bloomsbury Press).

Herspring, D.R., 2010. "Is Military Reform in Russia for 'Real'? Yes, But," ch. 2 in *The Russian Military Today and Tomorrow: Essays in Memory of Mary Fitzgerald*, edited by S.J. Blank and R. Weitz (Carlisle, PA: Strategic Studies Institute, U.S. Army War College), July, pp. 151-91.

Herspring, D.R., 2010. "Putin, Medvedev and the Russian Military," ch. 12 in *After Putin's Russia: Past Imperfect, Future Uncertain*, edited by S.K. Wegren and D.R. Herspring. Fourth Edition (Lanham, MD: Rowman & Littlefield), pp. 265-89.

Huntington, S.P., 1957. *The Soldier and the State: The Theory and Politics of Civil-Military Relations* (Cambridge, MA: Harvard University Press).

Janowitz, M., 1964. *The Professional Soldier: A Social and Political Portrait*. New York: The Free Press.

Junger, S., 2010. *War* (New York: Twelve—Hachette Book Group).

Kilcullen, D., 2010. *Counterinsurgency* (Oxford: Oxford University Press).

Kohn, R.H., 2008. "Coming Soon: A Crisis in Civil-Military Relations," *World Affairs*, Winter, http://www.worldaffairsjournal.org/articles/2008-Winter/full-civil-military.html/

Korb, L., 1976. *The Joint Chiefs of Staff: The First Twenty-Five Years* (Bloomington, IN: Indiana University Press).

Korb, L., 2010. "The public's disconnect with the military," March 27, http://www.signonsandiego.com/news/2011/mar/27/the-publics-disconnect-with-the-military/

Korb, L. and L. Conley, 2011. "War in an age of deficits," *CNNMoney*, March 23, http://money.cnn.clom/2011/03/23'/news/economy/Libya_cost_lawrence_korb_/index.htm/

Kurth, J., 2007. "Variations on the American Way of War," ch. 2 in *The Long War: A New History of U.S. National Security Policy Since World War II*, edited by A.J. Bacevich (New York: Columbia University Press), pp. 53-98.

Libicki, M.C., 2009. *Cyberdeterrence and Cyberwar* (Santa Monica, CA: RAND Corporation Project Air Force).

Martel, W.C., 2007. *Victory in War: Foundations of Modern Military Policy* (Cambridge: Cambridge University Press).

Millett, A.R., W. Murray, and K.H. Watman, 1988. "The Effectiveness of Military Organizations," ch. 1 in *Military Effectiveness, Volume I: The First World War*, edited by Millett and Murray (Boston, MA: Unwin, Hyman), pp. 1-30.

Moskos, C., J.A.Williams and D.R. Segal, 2000. *The Postmodern Military: Armed Forces After the Cold War* (New York: Oxford University Press).

Nagl, J.A., 2005. *Learning to Eat Soup with a Knife: Counterinsurgency Lessons from Malaya and Vietnam* (Chicago, IL: University of Chicago Press).

Oliker, O., K. Crane, L.H. Schwartz and C. Yusupov, 2009. *Russian Foreign Policy: Sources and Implication* (Santa Monica, CA: RAND Corporation).

Otterbein, K.F., 2009. *The Anthropology of War* (Long Grove, IL: Waveland Press).

Owens, M.T., 2011. *U.S. Civil-Military Relations after 9/11: Renegotiating the Civil-Military Bargain* (New York: Continuum Publishing Group).

Payne, J.L., "Deconstructing Nation Building," *The American Conservative*, October 24, 2005, pp. 13-15, reprinted in *International Politics: Enduring Concepts and Contemporary Issues*, edited by R.J. Art and R. Jervis (Boston, MA: Longman, 2001) pp. 445-9.

Perlmutter, A., 1977. *The Military and Politics in Modern Times: On Professionals, Praetorians, and Revolutionary Soldiers* (New Haven, CT: Yale University Press).

Petraeus, D.H., 2010-11. "Lessons of History and Lessons of Vietnam," *Parameters* (Winter), 40th Anniversary Issue, pp. 48-61.

Sarkesian, S.C., 1981. *Beyond the Battlefield: The New Military Professionalism* (New York: Pergamon Press).

Sarkesian, S.C. (ed.), 1980. *Combat Effectiveness: Cohesion, Stress, and the Volunteer Military* (Beverly Hills, CA: Sage Publications).

Sarkesian, S.C. and R.E. Connor, Jr., 2006. *The U.S. Military Profession into the Twenty-First Century: War, Peace and Politics*. Second Edition (New York: Routledge).

Shevtsova, L., 2007. *Russia – Lost in Transition: The Yeltsin and Putin Legacies* (Washington, D.C.: Carnegie Endowment for International Peace).

Shultz, R.H., Jr. and A.J. Dew, 2006. *Insurgents, Terrorists and Militias: The Warriors of Contemporary Combat* (New York: Columbia University Press).

Treverton, G.F., 2009. *Intelligence for an Age of Terror* (Cambridge: Cambridge University Press).

West, B., 2011. *The Wrong War: Grit, Strategy, and the Way Out of Afghanistan* (New York: Random House).

Wilson, I. III, 2007. *Thinking beyond War: Civil-Military Relations and Why America Fails to Win the Peace* (New York: Palgrave Macmillan).

Zinni, General T. and T. Koltz, 2006. *The Battle for Peace: A Frontline Vision of America's Power and Purpose* (New York: Palgrave Macmillan).

Index

For Product Safety Concerns and Information please contact our EU
representative GPSR@taylorandfrancis.com
Taylor & Francis Verlag GmbH, Kaufingerstraße 24, 80331 München, Germany